Studies in Logic

Volume 88

Belief Attitudes, Fine-Grained Hyperintensionality and Type-Theoretic Logic

Volume 68
Logic and Philosophy of Logic. Recent Trends in Latin America and Spain
Max A. Freund, Max Fernández de Castro and Marco Ruffino, eds.

Volume 79
Games Iteration Numbers. A Philosophical Introduction to Computability Theory
Luca M. Possati

Volume 80
Logics of Proofs and Justifications
Roman Kuznets and Thomas Studer

Volume 81
Factual and Plausible Reasoning
David Billington

Volume 82
Formal Logic: Classical Problems and Proofs
Luis M. Augusto

Volume 83
Reasoning: Games, Cognition, Logic
Mariusz Urbański, Tomasz Skura and Paweł Łupkowski, eds.

Volume 84
Witness Theory. Notes on λ-calculus and Logic
Adrian Rezuş

Volume 85
Reason to Dissent. Proceedings of the 3rd European Conference on Argumentation. Volume I
Catarina Dutilh Novaes, Henrike Jansen, Jan Albert van Laar and Bart Verheij, eds.

Volume 86
Reason to Dissent. Proceedings of the 3rd European Conference on Argumentation. Volume II
Catarina Dutilh Novaes, Henrike Jansen, Jan Albert van Laar and Bart Verheij, eds.

Volume 87
Reason to Dissent. Proceedings of the 3rd European Conference on Argumentation. Volume III
Catarina Dutilh Novaes, Henrike Jansen, Jan Albert van Laar and Bart Verheij, eds.

Volume 88
Belief Attitudes, Fine-Grained Hyperintensionality and Type-Theoretic Logic
Jiří Raclavský

Studies in Logic Series Editor
Dov Gabbay dov.gabbay@kcl.ac.uk

Belief Attitudes, Fine-Grained Hyperintensionality and Type-Theoretic Logic

Jiří Raclavský

ISBN 978-1-84890-334-0

College Publications
Scientific Director: Dov Gabbay
Managing Director: Jane Spurr

http://www.collegepublications.co.uk

Printed by Lightning Source, Milton Keynes, UK

Dedicated to my family and my friends for their love, friendship and support.

Preface

This book models belief attitudes/belief sentences within a type-theoretic (higher-order, multimodal) logic. First, let me put this intellectual enterprise on the map of the current research of our conceptual scheme, while I will also introduce some basic terminology and main ideas.

The crucial terminology employed in this book can be summarised in the following (informal) definition:

Definition 1 (Belief sentence, belief attitude)

A *belief sentence* typically takes the form "A believes that φ.", where "A" is an expression representing an *agent*, to whom the *belief attitude* is *ascribed* by a *speaker*; the verb such as "believe" presents the *belief operator*, and the *nested sentence* ('that'-clause) "that φ" presents the *object of the belief attitude* (or: *belief*).

Here are some obvious examples of belief (and similar) sentences:

"Xenia believes that the morning star is a planet."
"Xenia knows that $1 + 2 = 3$."
"Yannis does not assume that Xenia knows that the morning star is a planet."

Typical objects of beliefs are *'propositions'* (intuitively, they are meanings of declarative sentences)[1] which present facts, or at least hypotheses, concerning the external world of the believer or 'internal worlds' of (other) agents.

Belief attitudes are usually considered as the representative of a whole family of similar attitudes, most notably *knowing*, *thinking*, *refuting*, *asserting*, etc. Often, the term "*propositional attitude*" has been used as a neutral term. However, this expression, originally Russell's [355], does not seem to be comprehensive enough since so-called *de re* beliefs, frequently discussed in the field, are

[1] Some theoreticians explicate 'propositions' as propositions of possible world semantics. Propositions are also used in this book (thought for a purpose different from explication of sentential meanings), which is why I use single quotation marks for the intuitive non-technical notion.

not attitudes towards 'propositions', but towards individuals and 'attributes'. This is one of the reasons why I rather prefer the term "belief attitudes".

Since belief ascriptions and the reasoning involving them are frequent both in everyday life and science, one requires their *analysis*. The first such analyses already appeared at the birth of modern logic and analytic philosophy in the late 19th and early 20th century, and, since the 1950s, numerous rival proposals have been presented. Nowdays, belief sentences are studied in various aspects and details in the *philosophy of language* and in *epistemology*, in *formal epistemology* and in *epistemic logic* (*EL*), and also in *formal semantics*.[2] One may therefore agree with Anderson's emphasis of the problem of belief sentences/belief attitudes:

> the problem of the formal semantics of propositional attitudes is one of the hardest in philosophy.
>
> Anderson [8], pp. 338–339

The present book operates within the intersection of all the above-mentioned fields, while it mainly contributes to the investigation of various topics addressed by general EL and formal semantics. For such a logically-oriented investigation, an accent on form is typical and so even this book offers a *formal analysis* of belief.[3] Like other theoreticians such as Hintikka and Montague, by offering a formal analysis of belief sentences and some main rules governing their inferential behaviour, I will not yet propose the ultimate analysis of belief or knowledge.

According to the widely-adopted assumption, the objects of belief attitudes are 'propositions', not sentences, for 'propositions' are language-independent. But what are 'propositions'? 'Propositions' are theoretical entities that serve in many roles, not only as objects of our propositional attitudes, but also as the meanings of meaningful (declarative) sentences, as bearers of truth values, as components of logical arguments, etc. They have various qualities: they are possible or necessary, they are intrinsically bearers of truth-conditions, they are contents of assertions, they are compositional values of 'that'-clauses, they are the bearers of probabilities, etc. (see e.g. Pagin [285] for discussion). A well-elaborated logical theory of 'propositions' should treat their dominant properties and fulfil related theoretical demands. I am convinced that the current proposal provides a good example of such a theory.

---

2 The topic of propositional attitudes (alternatively: *belief reports*, *belief ascriptions*) is covered by an abundance of literature, see e.g. Richard [348], Anderson and Owens [9], Künne, Newen and Anduschus [223]; for an introduction, see e.g. Bäuerle and Cresswell [28], McKay and Nelson [249], Schwitzgebel [366].

3 Not a 'content' analysis, which is in the focus of general epistemology and philosophy; see e.g. Steup [391], Williamson [449].

The overall approach adopted in this book is a top-down one. The proposed logical system is not gradually enriched to capture more and more phenomena related to selected simple rock-bottom cases. Rather, I start with a comprehensive system applicable to a substantial fragment of natural language and focus on problems of belief sentences within it. This way one receives a sort of proof that the proposed analysis of belief sentences will not face some radical future revision. The approach also resonates with the call for elaborating EL, so fittingly tailored to sentences about knowledge and belief, into a wide-ranging logical picture of our language and reasoning:[4]

> I think two things will be necessary: epistemic logic should be incorporated into a logic that treats not just knowledge and belief, but also inference, understanding, and perhaps other propositional attitudes, and ties with more central parts of logic must be established—presumably proof theory and, more, generally, the theory of recursive functions.
>
> Anderson [8], pp. 338–339

I hope that I will take some very small steps forward in this enterprise.

The logical framework adopted in this book is a sort of *type theory*, a higher-order logic with multiple modalitites. One of the main arguments in favour of its employment is its great expressive power, and so its capability to control the validity of a vast range of inferences. The particular system I use is a substantial revision of the *partial type theory* developed by the Prague Spring 1968 refugee Pavel Tichý (*1936 Brno, the Czech Republic – †1994 Dunedin, New Zealand) in the 1970s and 1980s (see [419, 422]). It appertains to the famous system of intensional logic/semantics by Montague [258], but it is more akin to the current systems of *type-theoretic semantic* – see e.g. Ranta [342], Muskens [270], Francez [135], Chatzikyriakidis and Luo [62] – since it accentuates a proof-theoretic approach to meaning (despite that, the system still has a model-theoretic level).

Though there is a certain traditional mistrust of type theories among philosophers and some logicians, typed λ-languages and type theories have multiple successful uses both in formal/computational semantics and computer science (esp. functional programming, automatic theorem provers). Instead of philosophically arguing in favour of the legitimacy of the approach, I will rather highlight its versatility and develop its various applications within the main topic, with an accent on foundational issues.

It should be noted that the Tichýan approach owes a lot to Church's [68] *simple type theory*, but it significantly transcends it. Not only does it use both total and partial multiargument functions, its relaxed typing, reminiscent of

---

[4] Cf. also Scott's [367] famous advice on modal logic that recommends its reform onto multimodal logic.

Curry's type-polymorphism, increases its expressive capability. Moreover, it makes use of *ramification*, for which one may find an inspiration in works by Russell (e.g. [353]). Russell proposed the ramification of his intensional 'propositional functions' ('propositions' among them) to avoid circular *quantification* and the *semantic* and *epistemic paradoxes* (mainly the Russell-Myhill paradox) which results from an inadequate analysis of 'propositions'. Since 'propositions' are objects of 'propositional attitudes', the importance of their logically satisfactory analysis for our key topic is obvious, as aptly stated e.g. by Asher:[5]

> Quantification over propositions is a necessary component of any theory of attitudes capable of providing a semantics of attitude ascriptions and a sophisticated system of reasoning about attitudes.
>
> Asher [16], p. 11

This book follows Tichý in considering meanings to be his *constructions*, i.e. abstract, algorithmic computations of expressions' denotata, which he proposed in the mid-1970s (cf. [419]) and defended especially in [422]. This neo-Fregean proposal nicely fits the current demand for *fine-grained hyperintensional meanings*. Constructions are hyperintensions, since for any possible world intension (or extension) there exist infinitely many (congruent) constructions which construct it. Since constructions are structured (which is apparent e.g. from their records in λ-notation), they aptly play the role of fine-grained meanings.

* * *

Overview of the book content.

1. *Introduction: substitution in belief sentences and the hyperintensional conception of meaning.* First, I introduce the main notions and ideas assumed in the book, as they have been discussed in philosophical logic and the philosophy of language. A notional centre is the well-known Frege's paradox concerning the substitutivity of identicals, for it significantly affected our models of belief ascriptions. Among the various responses to the paradox, the influential possible world semantics recently met a criticism for its incapability to manage also hyperintensional contexts. Here I introduce the main idea of Tichý's approach: hyperintensions as meanings are abstract algorithmic computations (which fit the idea that meanings are fine-grained) that compute denotational values of terms, while the denotata are extensions or possible world intensions.

2. *Partial type theory.* In the second chapter, I provide a novel exposition and adjustment of Tichý's type theory (TTT). I offer a motivation for type theories

[5] It should be noted that the rules concerning quantification presuppose substitution, although they do not belong to the language of first-order logic, but its meta-language. In this book, however, substitution belongs to its main formal language and so I treat it explicitly (quantification is treated subsequently).

(*TT*) in general and briefly introduce their two main forms: simple and ramified TT. TTT is rather complex, since it is a ramified TT with a Church-like simple TT embedded in it. I gradually introduce a typed λ-calculus as a language for TTT, and offer a model-theoretic semantics for its terms (their proof-theoretic semantics is offered in the next chapter).

3. *Natural deduction for partial type logic.* For TTT, a natural deduction in sequent style exists; its crucial part was developed by Tichý in the mid-1970s, while Kuchyňka recently supplemented it by rules needed for TTT. The system is inevitably complex esp. because of partiality (the terms may lack a denotational value, functions may lack a value); it has more than thirty basic derivation rules. Similarly, as in typed λ-calculus, substitution plays an important role here because it occurs in some important derivation rules. In *Appendix A* I prove the generalisation of Tichý's Compensation Theorem that expresses correctness of substitution within TTT.

4. *Transparent hyperintensional logic.* TT is a versatile tool for the formalisation of natural language, since it uses a number of modal operators, their modifiers, etc. I introduce and elaborate Transparent hyperintensional logic (THL), a 'spin-off' of Tichý's original approach called Transparent intensional logic (TIL). THL's fundamental idea was suggested by Kuchyňka in 2016 and presented in my various papers; I offer here its most elaborated version. Both TIL and THL resemble the famous intensional logic by Montague in that they do not provide a logic in the narrow sense of axioms and rules, but rather a theory: a higher-order logic with a number of postulates that govern the behaviour of various extra-logical constants. Thus, THL unites both model-theoretic and proof-theoretic approaches to meaning, it is a type-theoretic semantics.

5. *Belief and substitution.* While Chapter 4 offers a formalisation of a basic fragment of language (singular terms, predicates and verbs, quantifying phrases, simple sentences, etc.), Chapter 5 exclusively focuses on belief sentences: belief *de dicto*/*de re*, nested beliefs, cross-reference in belief contexts, etc. I focus on substitution into belief sentences and also well-known examples of arguments that utilise it.

6. *Limits of belief and knowledge.* Chapter 6 focuses on two main themes. I mainly show how to translate systems of EL to the framework of THL, while I utilise our knowledge of substitution as described in Chapter 5. Epistemic capabilities of agents are modelled as limited, for one uses here a sort of ramified TT. However, the first part of the chapter discusses another kind of limitation: an agent's epistemic capability is limited because it is bound to use only the derivation rules it masters. This way I dismiss the objection that the present approach to belief sentences is too restrictive, while I am still capable of precluding the Paradox of Logical Omniscience.

7. *Belief and paradoxes.* Many paradoxes that affect various analyses of belief sentences are easily solved within the approach defended in this book. I

compare it with the TT-approach to semantic paradoxes (esp. the Liar paradox affecting the notion of truth), as already suggested by Russell. But I show its incapability to solve Church-Fitch's paradox of knowability because of the General Principle of Reducibility that holds in TTT. Moreover, I investigate principal limits inevitably put on the explication of notions such as truth, assertion, necessity and knowledge. An analogue of the Knower Paradox is constructed, while the original paradox is critically examined. In conclusion, I am forced to restrict the most basic rule concerning the notion of knowledge, the factivity rule.

* * *

Some linguistic agreements.

The term "*function*" usually means function in the purely extensional sense of mapping from arguments to values. Not in the sense of rule, for which I usually use the term " 'function' ". By function I do not mean a relation as a set consisting of ⟨argument, value⟩ couples, because TT usually takes the notion of function as primitive and the notion of set as derived (not the other way around, as is usual).

The term "*proposition*" always means a possible world proposition, i.e. a function from possible worlds (and perhaps also time instants) to truth values. The intuitive notion of sentential meaning is not called "proposition", but " 'proposition' ". 'Propositions' are explicated as certain Tichý's constructions. A similar agreement applies to the term "*intension*", which usually means possible world intension.

The term "*set*" is often used in its intuitive sense; in designated contexts, set is understood as a certain characteristic function. The term "*operator*" has a comparatively relaxed use: it often means a logical symbol such as $\rightarrow$ (though not in an exclusively syntactic sense), or its meaning, i.e. a certain Tichý's construction, not the set-theoretic entity (say a function) constructed by the construction. Double quotation marks are frequently used for quoting an expression. Single quotation marks indicate a meaning shift.

* * *

Intended reader, chapter dependence.

The book is intended for readers with at least undergraduate knowledge of philosophical and non-classical logics and related topics of the analytic philosophy of language and formal semantics. Some topics might be interesting for computer scientists, mathematicians and linguists. Chapters 2 and 3 are rather technical, they expose a system that is applied within the subsequent chapters. Chapter's dependence: $1 \longrightarrow 2 \longrightarrow \ldots \longrightarrow 7$; the chapters with greater numbers are relatively independent.

* * *

Thanks.

My sincere thanks go to many logicians and philosophers who contributed, in personal communication, to my investigation presented in this book, among others (lexicographical order): Anna Brożek, Massimiliano Carrara, Danielle Chiffi, Tadeusz Ciecierski, Jacques Dubucs, Duško Duždič, Chris Fox, Melvin Fitting, Pierdaniele Giaretta, Richard Goldblatt, Sten Lindstrøm, Roussanka Loukanova, Maria Manzano, Edwin Mares, José Martínez Fernández, Piotr Kulicki, Piotr Łupkowski, Marek Nasniewski, Graham Oddie, Marco Panza, Peter Pagin, Thomas Piecha, Francesca Poggiolesi, Graham Priest, Giuseppe Primiero, Ester Ramharter, Veikko Rantala, Ofra Rechter, Robert van Rooij, Hans Rott, Gabriel Sandu, Peter Schroeder-Heister, Gila Sher, Krzysztof Szymanek, Luca Tranchini, Robert Trypuz, Peter Verdée, Heinrich Wansing, Andrzej Wiśniewski, Dag Westerståhl, Zsofia Zvolensky. From the Czech and Slovak region I must mention Marta Bílková, Libor Běhounek, Petr Cintula, Pavel Cmorej, Ludmila Dostálová, Petr Hájek, Petr Koťátko, Petr Kuchyňka, Ladislav Kvasz, Ondrej Majer, Vilém Novák, Ivo Pezlar, Jiří Rosický, Igor Sedlár, Karel Šebela, Jan Štěpán, Vítězslav Švejdar, Jiří Zlatuška, Marián Zouhar, and many more. Special thanks go to Jindra and Veronica Tichý and the reviewers, who read the manuscript and suggested a number of improvements. Of course, none of them is responsible for any infelicities or errors possibly remaining in the present text.

I also thank the people who contributed to the current paradigm in logic and analytic philosophy, and the creators of LaTeX and KOMA-Script, the systems used for writing and typesetting this book. The author extends his thanks to the linguistic editors and also to Jan Štěpánek and Tereza Kunešová for help with the typesetting and bibliography. The author thanks the institutions for enabling the physical existence of the book, namely the Grant Agency of the Czech Republic (GAČR) (the grant no. GA16-19395S "Semantic Notions, Paradoxes and Hyperintensional Logic Based on Modern Type Theory"; Appendix and chapter 3 were elaborated or significantly expanded with support of the grant no. GA19-12420S "Hyperintensional Meaning, Type Theory and Logical Deduction") and the Department of Philosophy (Faculty of Arts) at Masaryk University.

Contents

Chapter 1

Introduction: substitution in belief sentences and the hyperintensional conception of meaning

1.1 Overview of the chapter

In this introductory chapter, I will frame the topics of the present book into a wider context of investigation of belief attitudes. I will partly focus on substitution for it displays the need for appropriate individuation of objects of beliefs, which was stressed by Frege when he encountered the famous *Frege's paradox* (1.2). Its investigation led to the development of intensional logic and possible world semantics (1.2.2), and then – via the *Paradox of Hyperintensional Context* – even hyperintensional logic and semantics (1.3, 1.4). I will briefly summarise its classical solutions and the recent development, esp. understanding hyperintensions as certain algorithms (1.4). In 1.4.1 and 1.4.2, I will informally introduce Tichý's notion of *construction*, i.e. an algorithmic computation, which will be considered as a model of meaning. We will see that Tichýan conception draws advantages from several leading approaches of the 20th century.

1.2 Frege's paradox

In his highly regarded paper "Über Sinn und Bedeutung", Frege [143] raised the important question whether it is always appropriate to *substitute* on the basis of *identity statements*, i.e., whether one can always apply *Leibniz's rule of substitution* ("*substitution of identicals*"):[1]

[1] The frequently quoted variant of Leibniz's formulation of the principle is *"Eadem sunt quorum unum potest substitui alteri salva veritate"*.

Definition 2 (Schema of Leibniz's rule of substitution, (SI))

$$\text{(SI)}\ \frac{\begin{array}{c}\varphi[t_1/x] \\ t_1 = t_2\end{array}}{\varphi[t_2/x]}$$

where φ is a certain formula, x is a variable, t_1 and t_2 are certain terms, "$=$" is the identity sign and "$[t_i/x]$" says that every occurrence of x in φ is substituted by t_i.

An application of (SI) is surely unproblematic in cases such as:

Example 1 (Valid argument licensed by (SI))

$$\frac{\begin{array}{c}\text{"The morning star is a planet."} \\ \text{"The morning star} = \text{the evening star."}\end{array}}{\text{"The evening star is a planet."}}\ \text{(SI)}$$

but Frege discovered that (SI) *fails* in cases such as this:

Example 2 (Invalid argument licensed by (SI))

$$\frac{\begin{array}{c}\text{"Xenia believes that the morning star is a planet."} \\ \text{"The morning star} = \text{the evening star."}\end{array}}{\text{"Xenia believes that the evening star is a planet."}}\ \text{(SI)?}$$

The argument is obviously invalid because its premises can be true while its conclusion is not: though the agent can believe that the morning star is a planet, she need not believe an equally interesting astronomical fact that the evening sky is also dominated by a planet.

The essence of *Frege's paradox* (or *Frege's puzzle*), often called the *Paradox of Identity*, is the fact that the apparently valid rule (SI) licences an apparently invalid argument.[2]

1.2.1 Classical solutions to Frege's paradox

In this section, I offer a summary of the classical approaches by Frege, Russell, Quine, Carnap and his followers to Frege's paradox.

---

[2] It matches with the now standard definition of paradox, see Sainsbury [358].

i. Frege's sense–denotation distinction

When seeking the source of the (SI) failure, Frege [143] rightly focussed on two problems: How identity statements of form $t_1 = t_2$ work?, What is the logical impact of 'indirect speech' (*oratio obliqua*) in the second type of arguments?

In his famous theory on *sense* ("Sinn") and *denotation* ("Bedeutung"), Frege claimed that expressions occurring in *extensional context* (he called it "gerade", i.e. "direct"; Quine: "transparent") serve for speaking about their *denotata*, i.e. about extensions. Extensional context can preliminarily be defined as a context which is not governed by an intensional or hyperintensional operator (verb) such as "believe". For example, the second premiss of Example 2 presents an extensional context. In its identity statement, both terms "the morning star" and "the evening star" serve for speaking about a denotatum, which is Venus. For such expressions, one can freely substitute *equivalent* (some say: *co-referential*) *expressions.*[3]

Frege claimed that indirect speech functions differently: its purpose is to point to the senses of expressions, which we now call (linguistic) *meanings*. In the first premiss and the conclusion of Example 2, the terms "the morning star" and "the evening star" occur in the indirect context ("ungerade", Quine: "oblique"), which is often called *intensional context*. The terms stand for senses which I will denote here by "THE MORNING STAR" and "THE EVENING STAR". Intensional context is governed by an intensional (below, we will see that even a hyperintensional) operator.

According to Frege, substitutions of co-denotative expressions, which do not have an identical meaning, are not properly licensed in such an intensional context, for it is not adequate to substitute expressions representing senses for expressions representing the corresponding denotata, and *vice versa*. In other words, expressions coming from extensional contexts cannot be substituted for expressions in intensional contexts.[4]

---

[3] In this book, instead of the term "equivalent", I will use the term "*v-congruent*" for a relation between models of meanings (namely between Tichý's constructions); the term "co-referential" will be reserved for a relation between expressions.

[4] This second formulation is not as precise as the first one since the notions of extensional and intensional contexts are not precisely defined; for the purpose of this book, they need not to be. I will only define hyperintensional contexts. Moreover, Russell first noted that a substitution into seemingly (hyper)intensional contexts is possible (see point iii. and 5.4.4 for more).

ii. Russell's elimination of descriptions

In his acclaimed study "On Denoting", Russell [354] rejected Frege's two-tier semantics, since he found the source of the (SI) failure in the inappropriate treatment of names. He split them in two categories, that of i. *proper names* (e.g. "Venus") and ii. (definite) *descriptions* (e.g. "the King of France" or the 'hidden' description "the morning star"), while he maintained that descriptions only contribute to sentential meaning by the meanings of predicates involved in them, so they are eliminable. The (schematic) sentence "The F is G.", for example, is thus analysed as $\exists x(F(x) \wedge G(x) \wedge \forall y(F(y) \rightarrow (x = y)))$.

While proper names create no problem when applying (SI), the failure of (SI) happens if descriptions are not eliminated from sentences in a proper fashion. In the first premiss (or the conclusion) of Example 1, only one proper elimination is possible. In Example 2, however, two eliminations are possible: they differ in scope of the quantifier $\exists$, which has a logical impact.

According to Russell's early work, the first premiss (as well as the conclusion) of Example 2 is a belief sentence, the object of the reported attitude being a 'proposition'. Such *Russellian 'structured propositions'* (which have been recently widely promoted, see e.g. Almog [4], Soames [379], and point v. of 1.3.1) are often represented using the symbolism of first-order logic. The two possible eliminations, which correspond to the a. *wide scope* (Russell: "*primary occurrence*") and to the b. *narrow scope* (Russell: "*secondary occurrence*") of the description in the whole sentence, can then be formalised as follows:

Example 3 (Wide and narrow scope of description)

a. $Bel_a \exists x(M(x) \wedge P(x) \wedge \forall y(M(y) \rightarrow (x = y)))$
– *narrow scope* of description

b. $\exists x Bel_a(M(x) \wedge P(x) \wedge \forall y(M(y) \rightarrow (x = y)))$
– *wide scope* of description

where Bel, a, M, P are analyses of the expressions "believe", "Xenia", "(be) the morning star", "(be a) planet". A substitution that licenses Example 2 is only possible on a wide-scope reading of belief sentences.[5] Hence, if read appropriately, Example 2 is a valid argument.

iii. Quine's *de dicto/de re* beliefs distinction

In his famous paper "Quantifiers and Propositional Attitudes", Quine [315] followed up on Russell's observation and offered various further considerations.

---

[5] Of course, before an application of (SI) to b.-readings of the sentences, one must apply the rule of $\exists$ elimination (this is not possible for the case of a.-readings of the sentences).

Quine described case a. as "*quantification within*", while case b. as "*quantification into*". This led him to initiate long-term research of belief attitudes in terms of quantification (see e.g. Kaplan [205]). Though not yet in this paper, Quine also introduced the following widely-adopted terminology for the a.- and b.-reading.[6]

Definition 3 (Belief attitudes *de dicto* and *de re*)

a′. *Belief attitude de dicto* (or *de dicto belief attitude*)
– an agent has an attitude towards the semantic content of the nested sentence, i.e. towards a 'proposition'.

b′. *Belief attitude de re* (or *de re belief attitude*)
– an agent has an attitude, *inter alia*, towards a certain thing, to which a proper name/description occurring in the belief sentence refers.

Example 4 (*De dicto* and *de re* belief sentences)

a′. "Xenia believes that the morning star is a planet."
– *belief sentence de dicto*

b′. "Xenia believes the morning star to be a planet."
– *belief sentence de re*

In case b′., the 'res' in question is Venus.

The *de dicto–de re distinction* is obviously highly relevant to the logical analysis of belief sentences. In case a′., the speaker who ascribes (using the belief sentence) a certain attitude to Xenia can exchange the means of reference to Venus from "the morning star" to "Venus" or "the evening star" *salva veritate*.

In case a′., however, the speaker cannot do this, i.e. the term "the morning star" is not substitutable by the term "Venus", because sentence a′. is true independently on the particular individual (if any) Xenia considers to be the morning star. Sentence a′. might be uttered either by Zoë, who had discussed with Xenia which object Xenia considers to be the morning star, or by Yannis who was told a′. by Zoë, while she has not revealed to him which particular object Xenia considers to be the morning star.

Quine [315] refuted the idea that the objects of belief attitudes *de dicto* would be 'propositions' in the sense of Russell's medadic *'propositional functions'* (see e.g. Mares [244] for introduction) since they did not satisfy the *Principle of Extensionality of Functions*,

$$\forall f \forall g(\, \forall x(f(x) = g(x)) \rightarrow (f = g)\,),$$

[6] For discussion, see e.g. the frequently quoted papers by Sosa [386] and Burge [52]. The *de dicto/de re* belief distinction originated during the Middle Ages.

as Russell and Whitehead [445] claimed. For Quine, such 'intensional entities' lacked acceptable criteria of individuation.

Consequently, Quine (e.g. [317]) propagated the eradication of *de dicto* belief sentences from scientific language. However, he was well aware of the fact that they are not fully eliminable, and so even he provided their analysis. He rendered them as reporting certain attitudes towards linguistic expressions as such (more precisely, towards expressions of formal notation):

"A 'believe-true' $\ulcorner\varphi\urcorner$.",

where the predicate "believe-true" is Quine's neologism.

The proposal, called *sententialism*, is generally deemed flawed, for there are convincing arguments against it. For example, sententialism is incompatible with our assumption that non-language believers exist: Fido the dog can well believe that there is meat in the fridge without having a 'believe-true' attitude towards the English sentence "There is meat in the fridge.". More seriously, Xenia can believe that the morning star is a planet without having the smallest idea about the way how to express her belief using a language unfamiliar to her. It is, therefore, not adequate to ascribe her the attitude 'believe-true' towards the German sentence "Das Morgenstern ist ein planet.", though one can truly ascribe her the attitude 'believe-true' towards the sentence "The morning star is a planet.".

In its original form, this *translational argument* says that if our analyses of belief sentences are translated into another language (recall that quoted sentences within Quine's paraphrases cannot be translated), the validity of an argument – which depends on the truth of belief ascriptions – must not vary. The argument was originally used against Carnap's [57] analysis of belief sentences by Church [73], cf. the Churchian counter-example in 1.3.1.

iv. Carnap's method of extension and intension

Though Frege (e.g. [143, 138, 141]) vividly defended senses as entities accessible to all speakers and, therefore, as language- and speaker-independent (i.e. abstract), which are structured and have a hierarchical composition, he did not propose a mathematical model for the notion of sense. This became one of the main contributions of Carnap.

In his influential book "Meaning and Necessity", Carnap [57] replaced Frege's original terms "Sinn" and "Bedeutung" by the traditional, yet appropriate, terms "*intension*" and "*extension*". His "Method of Extension and Intension" systematically associates an expression with an extension in extensional context and with an intension in intensional context. For example, the extension of "the morning star" is Venus, while its intension is the individual concept THE

MORNING STAR. (Among further extensions, one finds e.g. truth values, sets of individuals, sets of sets of individuals.)

During the 1950s and 1960s, Carnap's analysis of the notion of intension in terms of state-descriptions was completed in terms of *possible worlds*:[7]

Definition 4 (preliminary) (Possible world, intension, proposition)

Possible world is an object, which represents a maximal consistent set of (actual or potential) facts. *(Possible world) intension* is a function from possible worlds to objects of a certain type. *(Possible world) proposition* is an intension whose values are truth values.

Carnapian solution to Frege's paradox, which was accepted by many writers mentioned below – notably Montague and Hintikka, is actually Fregean in spirit, while its novelty consists in the analysis of the notion of sense in terms of mathematically well-defined (extensional) objects, namely possible world intensions. Within the approach, whole systems of natural language analysis have been developed, which I briefly discuss in the next section.

1.2.2 Possible world semantics, intensional and epistemic logic

The family of systems is usually called *possible world semantics*, or sometimes (with a bit different meaning) *intensional semantics* or *intensional logic.*[8]

Definition 5 (Possible world semantics, PWS)

Possible world semantics (*PWS*) is a semantics which ascribes PWS intensions at least to some expressions as their meanings.

According to PWS, a sentence stands for that proposition that contains those and only those possible worlds in which the sentence is true (in which it holds). For example, a logically true sentence holds in all possible worlds, and so it is paired with the only proposition true in all possible worlds.

As mentioned in Preface, the most famous system of PWS was elaborated by Montague [258], who applied it to the substantial fragment of natural language (see esp. his [262, 263, 257]). The system was developed from the logical point of view esp. by Gallin [145]; for its recent revision, see e.g. Muskens [270]. Similar approaches were also developed by others, Tichý [419] being an ex-

[7] For an introduction to the vast topic of possible worlds see e.g. Menzel [250]. For the notion of (possible world) proposition, see e.g. Stalnaker [389].

[8] For an introduction to intensional logic, see e.g. Fitting [128], Gamut [146], Anderson [6]. See also e.g. Marcus [243], Bressan [48], Fitting [127], de Rijke [349].

ample. Montague's approach achieved a great popularity among linguists;[9] in fact, he founded a new branch of formal linguistics.

In the 1970s, PWS provided a breakthrough analyses that could not be achieved within the Procrustean bed of first-order logic with identity: analyses of conditional statements (not discussed in this book at all) and analyses of belief sentences/attitudes (within the context of PWS, the term "propositional attitudes" is more frequent). Since there is a plenitude of possible worlds, say n, and so there are 2^n propositions, PWS has an abundance of various propositions that can be offered as meanings for various sentences, and so it is capable of offering the following, rather plausible, model of belief attitudes.

Definition 6 (Belief attitudes according to PWS)

According to PWS, *belief attitudes* are relations between individuals and propositions; the objects of belief attitudes are thus meanings of the nested sentences of belief sentences.

Consequently, PWS is capable of correctly evaluating of many arguments that cannot be correctly evaluated by the extensional semantics that considers sentential meanings to be simply truth values. If the objects of propositional attitudes were truth values, there would be only two equivalence sets of inter-substitutable sentences. The following argument would then be incorrectly evaluated as valid.

Example 5 (Invalid argument about belief and PWS)

$$\frac{\text{“Xenia believes that the morning star is a planet.”} \quad \text{“That the morning star is a planet = that the evening star is a planet.”}}{\text{“Xenia believes that the evening star is a planet.”}} \text{(SI)?}$$

Yet PWS can rightly diagnose the argument as invalid. According to PWS, the second premiss states an identity of truth values of the sentences flanking the identity sign with regards to the possible world of evaluation, but not the identity of propositions for which the two embedded sentences stand. Note that the approach of PWS is in fact Fregean: since the object of the belief reported by the first premiss is a particular proposition, one cannot substitute an expression representing another proposition (or even a truth value) for the nested sentence of the first premiss.

Such PWS style of analysis is also incorporated in the current descendants of Montague's approach, systems which are often called *type-theoretic semantics*, see e.g. the important works by Ranta [342], Muskens [270], Francez [135],

[9] See e.g. Janssen [189], Dowty, Wall and Peters [97] or Partee and Hendricks [289].

Chatzikyriakidis and Luo [62]. The systems are based on *type theory*, usually a modification of Church's [68]. Because of higher-order quantification captured within it, they are *higher-order logics*;[10] because of involving various modal operators, they present a sort of *(multi)modal logic.*[11]

PWS's treatment of belief sentences is also involved in standard systems of *epistemic logic* (*EL*), which was founded by Hintikka in his seminal work "Knowledge and Belief" [177].[12] Unlike the type-theoretic approaches, which have a great expressive power, systems of EL are often concentrated on logical analysis of a few selected *epistemic notions* such as *knowledge*, denoted "K", or *belief*, denoted "Bel" (or "B"). Nevertheless, the range of epistemic notions analysed by EL is steadily increasing, see e.g. Fagin, Halpern, Moses and Vardi [120], Meyer and Hoek [253], and the recent synoptic book von Ditmarsch, Hoek and Kooi [95]. I will return to EL's evaluation of arguments concerning belief in chapter 6, where I will also show how to translate various ELs to the type-theoretic framework adopted in this book.

1.3 From possible world intensional semantics to hyperintensionality

By treating sentential meanings as propositions, PWS is unquestionably capable of capturing the logical strength of sentences: the more possible worlds a(n assertoric) sentence sets apart, the stronger it is. Since propositions are assigned to them, sentences are classified in accordance with their logical strength. However, no proposition is associated with an exclusive sentence; equivalence sets of sentences exist, and so whole groups of non-identical sentences are classified as having the same meaning.

Here is an example of a couple of sentences denoting the same proposition:

"The morning star is a planet."
"The morning star is a planet and $1 + 2 = 3$."

Using elementary logical rules, one can easily derive many further sentences from the same equivalence set.

So although PWS treats sentential meanings as more fine-grained than the extensional semantics, which assigns one of the truth values as meaning to

[10] For introduction, see e.g. van Benthem and Doets [35], Väänänen [438].

[11] For an overview of higher-order modal logic, see e.g. Muskens [268]; for an overview of philosophical aspects of multimodal logic, see e.g. Smets and Velázquez-Quesada [376]; see also e.g. Williamson [450].

[12] For introduction to EL, see e.g. Égré [113], Meyer [251, 252], Hendricks and Symons [169, 170].

sentences, it is still a rather course-grained semantics: the above reference sentences surely differ in their meaning, yet PWS treats them as identical.

This PWS shortage entails, *inter alia*, that each of the infinity of true mathematical sentences is assigned the same proposition (namely, the one proposition that is true in all possible worlds) as their meaning. One of its negative effects is that the sentences are inter-substitutable within belief contexts, e.g.

Example 6 (Invalid argument licensed, according to PWS, by (SI))

$$\frac{\begin{array}{c}\text{“Xenia believes that } 1+2=3.\text{”}\\ \text{“}(1+2=3)=\forall x\forall y\forall z\forall n((x^n+y^n=z^n)\rightarrow(n<3)).\text{”}\end{array}}{\text{“Xenia believes } \forall x\forall y\forall z\forall n((x^n+y^n=z^n)\rightarrow(n<3)).\text{”}}\ (\text{SI})?$$

The second premiss of the argument contains Fermat's Last Theorem (*FLT*), which has, according to PWS, the same meaning as the trivial mathematical truth "$1 + 2 = 3$." occurring in the same premiss, so they are inter-substitutable.[13] But one can easily imagine a scenario in which Xenia does not believe FLT, though she believes $1 + 2 = 3$. The argument is thus intuitively invalid, despite PWS evaluates it as valid. Therefore, PWS fails as a tool for controlling an argument's validity.

Remark. After replacing "believe" by "knows" in this argument, one obtains an instance of the family of arguments that are jointly (and with some further assumptions) called the *Paradox of Omniscience.* The paradox shows that, upon the standard analysis of knowledge by means of PWS, agents are treated as having excessive epistemic abilities. See 6 for discussion.

From a logical point of view, the problem of PWS' substitution into belief sentences was aptly described by Cresswell [83] as the *Paradox of Hyperintensional Contexts.*[14]

> It is well known that it seems possible to have a situation in which there are two propositions p and q which are logically equivalent and yet are such that a person may believe the one but not the other. If we regard a proposition as a set of possible worlds then two logically equivalent propositions will be identical, and so if 'x believes that' is a genuine sentential functor, the situation described in the opening sentence could not arise. I call this the paradox of hyperintensional con-

[13] To avoid the objection that the main operator "=" of the identity statement operates on truth values, one may replace it with "≡" that operates on the meanings of sentences flanking its symbol.

[14] For a more popular form that affected the philosophy of language, see Cresswell's book aptly called "Structured Meanings" [85].

> texts. Hyperintensional contexts are simply contexts which do not respect logical equivalence.
>
> Cresswell [83], p. 25

Cresswell's definition of *hyperintensional context* is widely adopted:

Definition 7 (Hyperintensional context, hyperintensional operator)

Hyperintensional context is a context in which the substitution of (logically) equivalent expressions with non-identical meanings is not possible. The (formalisations of) verbs "believe" ("*Bel*"), "know" ("*K*"), etc. are called *hyperintensional operators.*

A model of meaning, which disallows the undesirable substitution of expressions to hyperintensional contexts is often called *hyperintension.* Specifically, but not always, hyperintension is a semantic object with a fine-grained structure (see below).

Remark. When understanding hyperintension as something that determines a denotatum, which is a certain extension (e.g., a truth value), hyperintension is de facto intension in the original sense of the word. Some writers, e.g. Fox and Lappin [134] and Moschovakis [265], speak about intensions even when solving the puzzles of hyperintensionality. Therefore, when an author refers to an "intension", one must be careful if she means a hyperintension or a PWS intension.

Outside of EL, the Paradox of Hyperintensional Contexts also stimulated the research in formal semantics. There was a demand for

1. *hyperintensional meanings.*

Using a particular example, any semantics is insufficiently fine-grained for the analysis of natural language, if it cannot capture the difference in meaning of expressions such as e.g. "equilateral triangle" and "equiangular triangle". Clearly, we need meanings that outnumber equivalence sets of expressions.

Meanings whose structures better correspond to analysed expressions have been searched for. Recall that, although PWS offered a certain meaning for each subexpression of sentences such as "Xenia believes that $1+2=3$.", these partial meanings do not form any hierarchical structure; e.g., one cannot take the meaning of the whole sentence and read from it meanings of its subexpressions. Therefore, one also seeks

2. *structured/fine-grained meanings*

which suitably correspond to the structure of analysed expressions. Here is a convenient quotation from Lewis, who first noticed the problem:[15]

[15] For recent discussion of fine-grained meanings, see e.g. Jespersen [194], King [208].

> intensions for sentences cannot be identified with meanings since differences in meaning – for instance, between tautologies – may not carry with them any difference in intension. The same goes for other [linguistic] categories, basic or derived. Differences in intension, we may say, give us *coarse* differences in meaning. For *fine* differences in meaning we must look to the analysis of a compound into constituents and to the intensions of the several constituents.
>
> Lewis [235], p. 31

1.3.1 Some solutions to hyperintensionality

Various solutions to hyperintensionality have been offered; the following overview is far from exhaustive. Not all of these solutions solve both the problem of substitutability of expressions and structuredness of meanings. Although each can be criticised for some of its feature, each of the solutions has its own merit. The present book does not attempt to argue which approach is the 'best one', but rather to develop one of them.

i. A part of EL's research focussed on the limitation of excessive inferential consequences that result from the PWS analysis of knowledge (which I will discuss from another viewpoint in 6.2). Hintikka [175] proposed the well-known solution to the problem of logical omniscience, which revises PWS by adopting propositions that contain, besides standard possible worlds, even *impossible possible worlds* (for recent development and defence, see e.g. Jago [188]). In these worlds, it holds e.g. $1+2=3$ but not FLT; such worlds are thus appropriate for modelling the belief of an agent who believes $1+2=3$, but not FLT.

Various objections and worries against such a proposal can be raised (see e.g. Bjerring [43] and 6.2). One of them is principal from the viewpoint of this book: the problem of insufficient structuredness of PWS meanings is not addressed by the approach at all. For example, the sentences "The morning star is a planet." and "The morning star is a planet or $1+2=3$." have, according to PWS, certain logically independent propositions as their meanings, despite the fact that the intuitive meaning of the first sentence is a part of the meaning of the second sentence, being thus logically dependent.

ii. Such an objection can perhaps also be raised against the well-known proposal by Thomason [405], who utilised the framework of PWS. However, since Thomason's proposal is more sophisticated than I can show here, I rather speak about the Thomasonian proposal (the analysis is borrowed from Raclavský [328, 324]). In the Thomasonian proposal, each sentence from a certain equivalence set denotes the same proposition, say P, but each of them is distinguished by a specific 'mark', which is a primitive object π_i of a new domain used in the corresponding PWS, so sentential meaning is identified with $\langle P, \pi_i \rangle$.

But one may raise the following concern. Consider e.g. the sentences

(i) "The morning star is a planet."

(ii) "The morning star is a planet and $1 + 2 = 3$."

(iii) "The morning star is a planet and $2 + 1 = 3$."

(iii) follows from (i) and (ii), (ii) follows from (i) and (iii) because of the definition of disjunction, commutativity of addition and also the fact that the meaning of (i) is a part of both (ii) and (iii). In the Thomasonian approach, however, the couples $\langle P, \pi_1 \rangle$, $\langle P, \pi_2 \rangle$ and $\langle P, \pi_3 \rangle$ that represent the semantic content of (i)–(iii) harbour no such information. Although the sentences are treated as similar because their meanings encapsulate P, they are otherwise semantically dissimilar. It is then not clear how to explain which sentence entails which.

iii. Another way how to preserve PWS and capture the structuredness of meanings is a certain form of sententialism. One of its first variants was published by Lewis [235], who hanged extensions and intensions of an expression on a (mathematical) tree. Similar proposals were published by Larson and Ludlow [230], Richard [348] and (admittedly) Cresswell and von Stechow [86] (for whom meanings are n-tuples that contain e.g. the symbol λ coming from the formal representation of analysed expressions). All these proposals can effectively be criticised by the aforementioned translational argument, see Asher [17] (the paper is aptly called "Meanings Don't Grow on Trees"), Soames [378] and Tichý [421].

iv. Carnap [57] suggested a seemingly similar solution: an agent believes the term t, if she virtually believes every so-called *intensionally isomorphic* term. Two terms are intensionally isomorphic if their subexpressions express the same intensions in an isomorph arrangement. The definition of an intensional isomorphism enabled Carnap to avoid sententialism, which hinges on one selected notation (language). His proposal seems to be notationally independent, since the objects of attitudes have 'trans-notational status', so to speak.

Yet in his criticism, Church [73] developed a counter-example using intertranslatable, but intensionally unisomorphic expressions "zwei Woche" ("two weeks" in German) and "fortnight" (translatable to German only as "zwei Woche"). I adapt Church's example to also cover pure sententialism:

Example 7 (Counter-example to sententialism/intensional isomorphism)

a. "Xenia believes that Yannis has spent a fortnight in Spain."
b. "Xenia believes-true 'Yannis has spent a fortnight in Spain.'."
a′. "Xenia glaubt das Yannis habt zwei Woche im Spanien verbringt."
b′. "Xenia glaubt-wahr 'Yannis has spent a fortnight in Spain.'."

Now sentence a. entails b. (assume that for the sake of argument), a. is inter-translatable with a′., b. is inter-translatable with b′. But a′. is not translatable with b′. Moreover, a′. does not entail b′. because Xenia – mastering only German, her mother-tongue – need not understand English, and so she would deny that she has a 'believe-true' attitude towards the string of English words mentioned in b′. Moreover, with rudimentary knowledge of English, Xenia may affirm the sentence "Yannis has spent two weeks in Spain." without affirming "Yannis has spent a fortnight in Spain.".

As pointed out by Tichý [422], even when abstracting from such counter-examples, it is not clear how to explain the inter-translatability or intensional isomorphism of expressions that are equivalent, but not identical, without referring to hyperintensions as objects that mediate the denoted intension/extension.[16]

v. Recently, *'structured propositions'* have been largely discussed. Among the pioneer proposals, one can find e.g. Cresswell [85], Soames [380, 379, 383], Salmon [360], King [211, 210, 209], while most of them defended a broadly Russellian approach. Some authors, e.g. Soames [383], argue against propositions as a set of circumstances ('situations'), so-called *circumstancialism.*[17]

The structure of 'structured propositions' has been intensively discussed recently. For example, Soames [382] pointed out that the structure of (say) 'Othello loves Desdemona' cannot be identified with couples such as ⟨loving, ⟨Othello, Desdemona⟩⟩, since ⟨⟨Othello, Desdemona⟩, loving⟩ also seems to be a good candidate. For further arguments and discussion, see e.g. Keller [207], Ostertag [283], Pagin [285]. Many of the authors argue in favour of fine-grained meanings, see e.g. Jespersen [192, 194], King [208]. The problem of *propositional unity* is also studied, see e.g. Gaskin [150], Jespersen [194], García-Carpintero and Jespersen [147], Eklund [115].

vi. There are also other proposals, which differ in what they consider as hyperintension. Their essential feature is that they postulate a new category of

[16] For further discussion, see e.g. Thomason [407] and Cresswell [84].

[17] For criticism of Soames's argumentation, see e.g. Edelberg [112], Elbourne [116]). For recent defence of circumstantialism, see e.g. Elbourne [116], Berto [40], Ripley [350], Jago [187], who usually utilised impossible possible worlds. Various arguments in favour of the approach occurred in Stalnaker [390], Barwise and Perry [27], see also e.g. Faroldi [122].

entities that fulfil the role of hyperintensions. Many of their authors integrated PWS in their hyperintensional semantics, and often speak about *fine-grained intensionality*. See e.g. Bealer [30, 29], Zalta [453, 452], Fox and Lappin [134, 229], Chierchia and Turner [64], Turner [433], Pollard [304, 303], Leitgeb [232], Sedlár [370]. In this group, one can also find a specific variant that is quite unambiguous about the individuation of hyperintensions as certain algorithms. See the next two sections.

1.4 Neo-Fregean algorithmic semantics

First, I introduce another desideratum on a model of meaning:

> 3. *meaning has the nature of an algorithm* that computes the denotatum of an expression.

Remark. The term "algorithm" is not used here in the precise technical sense.

Point 3 patently adverts to Frege's sense–denotation distinction; I will thus speak about *neo-Fregean algorithmic semantics*.[18]

Definition 8 (Neo-Fregean algorithmic semantics)

Neo-Fregean algorithmic semantics offers a model of meaning which is 1. hyperintensional, 2. fine-greained 3. algorithmic.

For an elaboration and defense of such a semantics, see e.g. Muskens [271] and the authors mentioned in the rest of this section.

Recently, desideratum 3 has been defended by Girard [156] in his book "Types and Proofs". Girard has pointed out the fact that extensional content of expressions, which is offered by (so-called) denotational semantics, does not entirely capture the semantic content of expressions.

> we have an *equality*
>
> $$27 \times 37 = 999$$
>
> This equality makes sense in the mainstream by saying that the two sides *denote* the same integer and that $\times$ is a *function* in the Cantorian sense of a graph.
>
> This is the denotational aspect, which is undoubtedly correct, but it misses the essential point:

[18] Neo-Fregeanism was defended by various philosophers, see e.g. Forbes [133] and Chalmers [61]. For discussion of Fregeanism with regard to hyperintensionality, see Skipper and Bjerring [374].

> There is a finite *computation* process which shows that the denotations are equal. ... The two expressions have different *senses* and we must *do* something (make a proof or calculation ...) to show that these two *senses* have the same *denotation*
>
> Girard [156], pp. 1–2

Also, the acclaimed formal semantics elaborated by Moschovakis [266, 265] implements an algorithmic version of Frege's sense–denotation distinction:

> *algorithm which computes the truth value of* [sentence] χ ...
> This algorithm is the *referential intension* or just *intension* [bold face suppressed] of χ ...
> and we propose to model the sense of χ by its referential intension.
>
> Moschovakis [266], p. 211

Remarks. By writing about intensions, Moschovakis seems to follow up on the distinction between *intensional/extensional understanding of functions* and the distinction *intensional/extensional understanding of λ-terms.* In common mathematical understanding, which can be found in Bernoulli, Euler, but even Frege or Russell, a function is a certain 'expression' consisting of numbers (or numerals) and variables that 'computes' various values depending on numbers assigned to the variables.[19] In newer writings, it is common to write about functions as *rules* that prescribe how to transform input numbers into output values. Within a set theory that was widely accepted in logic, however, an extensional understanding of a function has been accepted: a function is a certain association between members of the domain of arguments and the members of the domain of values, i.e. it is a function as a *mapping* ("graph"). For each such extensionally individuated function, there exists an infinity of equivalent functions as rules.[20] In accordance with the extensional paradigm, Church (e.g. [77]), who founded λ-calculus, explained λ-terms (which code functions, or applications of functions to arguments) in the extensional way, though one may aptly understand them in the intensional way as procedures of arriving at certain extensional objects.

In this book, I will utilise the system by Tichý who proposed a neo-Fregean semantics as early as in his paper titled "Intensions in Terms of Turing Machines" [417] (its Czech version: [420]; the term "procedure" was then used for algorithm as a system of instructions):

---

[19] While Russell (e.g. [445]) wilfully accepted the intensional understanding of functions, the modern notion of functions as course-of-values appear sometimes in Frege's work. This led to certain ambiguities in his conception, since functions are also unsaturated entities that contain gaps; see e.g. Benis-Sinaceur, Panza and Sandu [34] or Tichý [422] for discussion.

[20] Some writers explicitly give notice about the two approaches. For example, Hindley and Seldin [174] distinguish between functions as mappings and functions as operations (which they represent by λ-terms or terms of combinatory logic).

> The sense of an expression is an entity linking the expression with its denotation. ... For what does it mean to understand, i.e. to know the sense of an expression? It does not mean actually to know its denotation but to know how the denotation can be found ... to know a method or procedure by means of which [a denotatum] can be identified
>
> Tichý [417], pp. 8–9

Moschovakis [266] realised that meanings need not have entirely all of the properties algorithms have, and so he specified them in terms of acyclic recursion (see e.g. Fitting [128] for more). Tichý – who called his final notion "*construction*" – aptly explained why the pure notion of the algorithm is not suitable:

> the notion of construction just defined is not to be confused with that of an algorithm. It is true that the two notions overlap: the construction $\text{Comp}^2(9 - 2)$, consisting in subtracting two from nine, is a simple algorithm for generating the number seven. This, however, is a degenerate sort of algorithm. A typical numerical algorithm is a procedure applicable to any of a *class* of input numbers, and the output number it generates depends on which input number it is applied to. But not only that; the very sequence of steps leading to the output may differ from one input to another. A construction is thus correlative not with the notion of algorithm itself but which what is known as a particular algorithmic *computation*, the sequence of steps prescribed by the algorithm when it is applied to a particular input. ...
>
> But not every constructions is an algorithmic computation. An algorithmic computation is a sequence of *effective* steps, steps which consist in subjecting a manageable object (usually a symbol a finite string of symbols) to a feasible operation. A construction, on the other hand, may involve steps which are not of this sort. The application of any function to any argument, for example, counts as a legitimate constructional step; it is not required that the argument be finite or the function effective. Neither is it required that the function constructed by a closure have a finite domain or to be effective. As distinct from an algorithmic computation, a constructions is an *ideal* procedure, not necessarily a mechanical routine for a clerk or a computing machine.
>
> Tichý [411], p. 526

1.4.1 Tichý's notion of construction

In Tichý's approach, the notion of construction is crucial. From the above quotation, one easily derives:

Definition 9 (preliminary) (Construction)

Construction is a (not necessarily effective) algorithmic computation. Constructions *construct* objects.

Comments.

i. It seems natural to understand constructions in a realistic way as language independent, abstract entities.[21] By meanings of certain expressions $E_1, E_2, ...$ one would then consider constructions expressed by them. For the sake of our communication, the construction is aptly recorded by means of λ-notation, since it conveniently reflects the structure of constructions (see chapter 2). In the nominalistic approach, on the other hand, construction would be identified with a certain 'canonical expression' – best with the λ-term or its tree representation[22] – and the analysis of the meanings of $E_1, E_2, ...$ would consist in the presentation of this 'canonical expression'. Against such a nominalistic approach, which is kindred to sententialism, one can raise e.g. the objections mentioned above; for this reason, the realistic approach is advisable.

ii. Constructions are understood as independent of anybody who contemplates them: they are not mental entities. Constructions are also understood as spatially and temporally abstract. Constructing is thus not a process located in space and/or time. For that reason, there is no difference between the notions of construction and their constructing.

Although it is not the purpose of these remarks to offer a proper comparison of Tichý's notion of construction with the intuitionistic (constructivistic) notion, mentioning at least basic similarities and dissimilarities is desirable. Because of their abstractness, Tichý's constructions obviously differ from the mentalistic notion of early intuitionists (e.g. Brouwer [50, 51]). On the other hand, contemporary intuitionists (see e.g. Howard [181], Sundholm [394] or van Dalen [91]) seem to consider a similar, abstract notion of construction. An obvious difference between Tichý's constructions and the intuitionistic ones is then the fact that the former ones do not have to be effective.

iii. Similarly as Frege [143], who considered different senses corresponding to the intersection of different medians of an equilateral triangle, Tichý used a geometrical example to explain his notion of construction in his paper titled "Constructions". He exposed it alongside his arithmetical example:

> I shall call such procedures *constructions*, borrowing the term from geometry, where we speak of various constructions of, say, the center of a circle, using rule and compass. ...
>
> The term '9 − 2' names the number seven. It does not name it, however, in the same *way* as does either '7' or '3 + 4'. It names it *qua* what results when two is subtracted from nine. Thus, apart from naming seven, '9 − 2' also expresses a specific indirect way of arriving at seven; '3 + 4' expresses a completely different way of arriving at the same number. Also '7' can be regarded as presenting seven *qua* the result of a particular, albeit trivial, procedure: the procedure consisting in starting with seven and leaving it at that. ...

[21] Since the defence of realism is not a topic of this book, it is only considered here as a useful assumption.

[22] For such a presentation of his notion of construction, see Fletcher [130].

> It is these numbers [namely 9 and 2] and the function [namely −] themselves that are involved in the construction of seven (just like points, line segments, and curves are involved in a construction of the center point of a circle)
>
> Tichý [411], p. 514

iv. The last (as well as the next) quotation speaks about the way in which constructions fulfil the idea of structuredness and fine-grained individuation: constructions are hierarchical complexes whose parts correspond to the structure of analysed expressions.

> Arithmetical expressions represent, or depict, constructions. The construction consisting in multiplying two by itself and subtracting three from the result, for example, finds its linguistic representation in the term '(2.2) − 3'. The primitive symbols '2', '3', '.', and '−' represent the primitive constituents of the construction (namely the numbers two and three and the operations of multiplication and subtraction respectively) and the way the symbols are arranged into the term is exactly parallel to the way those numbers and operations are organized into the construction.
>
> Tichý [422], p. 1

v. Finally, constructions seem to provide a solution to Frege's problem of logical adhesion (see Tichý [422]). They explain what binds a function and an argument into a functional whole. I.e., which way the meanings of subexpressions are combined into a compact complex of the whole expression's meaning.

Though constructions are representable e.g. by means of relations and their arguments, or by means of n-tuples, every such set-theoretical representation of constructions is only partial; constructions are not reducible to these notions. This can easily be seen in the case of relations: for an explanation of the bond that ties relation R to argument A, one needs, if not utilising the notion of construction, a certain relation. Yet another relation is required for the explanation of its bond to R and A, which leads to the emergence of an infinite regress.

1.4.2 Constructions as (linguistic) meanings

As repeatedly suggested above, properties of constructions make constructions good models for (linguistic) meaning. Exact details will be shown after the precise specification of the notion of construction in the next chapter, namely in chapter 4, where I expose and develop *Transparent hyperintensional logic* (THL), proposed by Kuchyňka (unpublished) as a modification of Tichý's original system called *Transparent intensional logic* (TIL). Various writers worked on Tichý's original system, e.g. Duží, Jespersen and Materna [110] or Raclavský

Kuchyňka and Pezlar [339].[23] For recent defences of Tichý's approach to meaning in the context of contemporary debates on 'structured propositions' and fine-grained meanings, see e.g. Jespersen [191] and Duží [104].

The convenience of constructions as (linguistic) meanings was already observed by Tichý in the mid-1970s, cf. Tichý [419]. Here are two quotations:

> The construction expressed is, in fact, what the linguisticians seem to be groping for they speak of the "deep structure" of a piece of language. In a word, it is the construction that a linguistic expression *means*, and that is depicted by the syntactic structure of the expression.
>
> Tichý [411], p. 515

> Constructions must be what we talk about and what expressions through which we communicate stand for ... An expression is simply a name of the construction depicted by it.
>
> Tichý [422], p. 224

Remark. Tichý [411] added that *understanding* an expression amounts to grasping the construction expressed by the expression (in a given language); *translating* an expression into another language amounts to expressing the same construction by an expression of the target language.

Since constructions are objects of type theory and the *semantics* is *type-theoretic*, it would be convenient to show how it fulfils the recently accentuated idea of the *proof-theoretic approach to meaning.* I will return to the topic in chapter 4, this is only the core idea.

The proof-theoretic approach to meaning was initiated by Gentzen's [152] observation that inferential behaviour of e.g. "$\wedge$" conforms to the rules (where "φ" and "ψ" are formulas and "/" derivability sign)

$$\wedge\text{-I}\ \frac{\varphi \qquad \psi}{\varphi \wedge \psi} \qquad\qquad \wedge\text{-E}_1\ \frac{\varphi \wedge \psi}{\varphi} \qquad\qquad \wedge\text{-E}_2\ \frac{\varphi \wedge \psi}{\psi}$$

The rules can be rewritten into the familiar truth table (i.e. the denotational semantics for "$\wedge$"), which means that the rules alone specify the semantics of "$\wedge$".

Type-theoretic semantics assumes that extra-logical constants obey the same proof-theoretic rendering of meaning. To illustrate, consider a definition of the meaning of the operator which should occur in lieu of "$\star$" by means of the following couple of rules:[24]

---

[23] TIL is also used in Oddie's [277] book about truth-likeness of scientific theories.

[24] The example is adapted from Oddie's and Tichý's [278] brief explanation of definitions as derivation rules.

$$\frac{n_1 + n_2 = n_3}{n_3 \star n_2 = n_1} \qquad \frac{n_3 \star n_2 = n_1}{n_1 + n_2 = n_3}$$

where the values of each n_i are natural numbers from $\mathbb{N}$, on which $=, +$ and the operator in lieu of "$\star$" operate. The operator "$-$", which should occur in lieu of "$\star$", is uniquely determined, so as its denotatum is determined by its (simple) meaning. The denotatum is clearly the subtraction function (as mapping), for no other function satisfies the constraints imposed by the rules.

In this book, I will look for rules that govern the operator "believe" (i.e. "*Bel*"), etc., as it occurs in various inferences. I will follow a number of requirements, many of which I have stated above.

Chapter 2

Partial type theory

2.1 Overview of the chapter

This extensive chapter introduces my adjustment of *Tichý's type theory* (briefly: TTT), a logical system involving a certain formal language (with its intended semantics) and a suitable deductive system.

To place TTT properly into context, I will describe main ideas of *type theories* (*TT*s), which can also be considered as typed λ-calculus (2.2). I will add remarks concerning the *Vicious Circle Principle(s)* (*VCP*; 2.2.5), as well as remarks on partial functions (2.4.3) and improper (abortive) constructions (2.3.2) that are integrated in TTT.

I will then introduce formal languages (2.3,2.5), the last of which is the *language for* TTT (briefly: $\mathcal{L}_{\mathsf{TTT}}$). As with other languages for TTs, $\mathcal{L}_{\mathsf{TTT}}$ is a specifically extended typed λ-calculus as language. I will state the ideas that motivate the transition from the simplest possible language to $\mathcal{L}_{\mathsf{TTT}}$. The proper *definition of* TTT (2.4.4) will be supplemented by examples (2.5.1) and a discussion of the interesting properties contained therein.[1]

Constructions will be systematically introduced (2.6) as direct semantic values of $\mathcal{L}_{\mathsf{TTT}}$'s terms. The objects constructed by the constructions are treated as their denotational semantic values. The semantics thus has two levels, it is neo-Fregean. I will add examples and comments (2.6.3), as well as definitions of the notions of subconstruction (2.6.1) and free variables (2.6.2). The deduction system for TTT will be introduced in the next chapter (3).

[1] My exposition of TTT in the style 'syntax first, semantics second' contradicts Tichý's [422] strict policy of offering exclusive 'semantics'. Though one may agree with Tichý's objections against the logical paradigm within which the method 'syntax first, semantics second' sometimes led to errors and misconceptions, the method was introduced to avoid even greater errors. Among others, the method presents logical things in an easily comprehensible way and enables a more efficient check of typos and other mistakes.

2.2 Main ideas of type theories

2.2.1 Simple type theory

One usually understands TT to be a system that consists of a. a definition of types (the proper part of TT), b. a typed λ-calculus (as language), c. a deduction system (natural deduction or sequent calculus). TT is based on the idea that it does not make much sense to apply a function to itself. This observation was described by Russell in his Appendix B "The Doctrine of Types" [356]. In other words, one cannot give a coherent interpretation to the expression "$f(f)$", where "f" is the name of a 'function'. Some writers, e.g. Frege [137] and Schröder (see Church [76]), anticipated Russell's findings by distinguishing levels of 'functions', which precludes the problem: to a certain 'function' f one can only apply a 'function' of a higher level; 'functions' are thus classified into pairwise disjoint 'levels'. Yet Russell's considerations were more in-depth. For the history of TTs, see e.g. Kamareddine, Laan and Nederpelt [202, 200].

In his Appendix B. [356], Russell proposed the first *simple TT* (*STT*). It enables the avoidance of the landmark *Russell's paradox* (Russell [356], see e.g. Link [238] for discussion), for the definiens of 'Russell's set' "$\{x : x \notin x\}$" is not a well-formed expression in STT. Russell later abandoned his STT in favour of his ramified TT (see 2.2.4 below), which was not, however, widely accepted. In the first half of the 20th century, its critics – Ramsey [341], Chwistek [429] and others – suggested various STTs. The acclaimed STT was proposed by Church [68]; here is a slight adaptation of his definition:

Definition 10 (Church's STT)

1. The symbols "o" and "ι" are *types*.
2. If "τ_1" and "τ_2" are arbitrary types, then the term "$(\tau_2\tau_1)$" is also a *type*.
3. Nothing is a *type*, unless it so follows from 1. or 2.

Its intended semantics: "τ_1" and "τ_2" denote certain sets of objects and "$(\tau_2\tau_1)$" denotes a set of functions from τ_1 to τ_2 (i.e. $\tau_1 \mapsto \tau_2$); types "ι" and "o" denote the set of individuals and the (two-membered) set of truth values, respectively. With the help of types, one 'decorates' expressions of formal language, and so e.g. "$f^{(\tau_2\tau_1)}(a^{\tau_1})$" becomes its well-formed term, while "$f^{(\tau_2\tau_1)}(f^{(\tau_2\tau_1)})$", which is a typed variant of "$f(f)$", is not type-theoretically correct and is therefore forbidden.

Though it is now common to view types as certain expressions, I will follow Tichý (and early Russell and even Church) in considering types as the sets denoted by the corresponding type terms:

Convention 1 (Type, type term)

Type is a collection (i.e. a set) of objects. Types are denoted by *type terms*.

In TT, three important facts are related to types:

(a) Each term of a (formal) language denotes an object of a particular type.

(b) Types serve as domains and ranges of functions.

(c) Types serve as ranges of variables.

With regards to (a), see 2.4.4. With regards to (b) and (c), the adoption of TT by Russell and other writers was often motivated by an attempt to avoid paradoxes. For example, the idea of a type classification of function ranges prevents 'Russell's set' (which is now understood extensionally as a certain characteristic function) to be in the range of its applicability. Russell's paradox is thus precluded. As a result of (b), variables contained in 'functions' (which are now understood in an intensional sense) cannot have unrestricted ranges; they are linked to a certain 'safe' domain. For example, the 'function' $\in$, which is involved in the definition of the 'Russell's set', applies to a variable tied up to a certain range within which $\in$ (as well as anything involving it) is not permitted to occur. Russell's paradox is thus precluded even in this understanding of its key ingredients.

TT received an unanimous acceptance in the field of computer science. Since the application of a function to a wrong type of input data would lead to an error (program crash, cycling), type safety has been enforced in most programming languages (Ada, C, C++, C♯, Java, Standard ML, Pascal, Common Lisp, Haskell, etc.).

2.2.2 Typed λ-calculus

The prevention of paradoxes within TT led to the adjustment of language in such a way that it is clear which object the term denotes. A convenient historical example is Church's STT, arising in reaction to the existence of the *Kleene-Rosser paradox* (Kleene and Rosser [213]), which occurs in Church's [70] original *untyped λ-calculus* (and also in Curry's [89] early combinatory logic). Church used the symbolism of his STT for typing the terms of his untyped λ-calculus, which then led to *typed λ-calculus* that was immune from the paradox. The system is often simply called "language of STT", or just "STT".

A language of λ-calculus *typed in Church style* ("Church-typing", "à la Church", or "explicit typing"), $\mathcal{L}_{\lambda_{Ch}}$, can be introduced using Backus-Naur form (*BNF*), which replaces a 'verbose' inductive definition.[2]

Definition 11 (Typed λ-calculus à Church, $\mathcal{L}_{\lambda_{Ch}}$)

$t^{\tau} ::= \mathrm{c}^{\tau} \mid x^{\tau} \mid [t_2^{\tau_2}\, t_1^{\tau_1}] \mid [\lambda x^{\tau_1}.t^{\tau_2}]^{(\tau_2\tau_1)}$

Remark. "c^{τ}" is a *constant*, "x^{τ}" is a *variable*, "$[t_2^{\tau_2}\, t_1^{\tau_1}]$" is called "*application*" and "$[\lambda x^{\tau_1}.t_1^{\tau_2}]$" is called "$\lambda$-*abstraction*".

Comments.

i. λ-calculus is a highly expressive language that is capable of embracing any mathematical expression. The term "$[t_2^{\tau_2}\, t_1^{\tau_1}]$" presents an application of the function represented by "$t_2^{\tau_2}$" to the argument represented by "$t_1^{\tau_1}$". The expression of the form "$[t_2^{\tau_2}\, t_1^{\tau_1}]$" is thus synonymous with the familiar mathematical expression of the form "$t_2^{\tau_2}(t_1^{\tau_1})$". For example, "$+(x, 1)$" can be rewritten in λ-calculus as "$[+\, x\, 1]$" (type symbols are now suppressed), or in the infix manner (which I will employ) as "$[x + 1]$".

The expression "$[\lambda x^{\tau_1}.t^{\tau_2}]^{(\tau_2\tau_1)}$" represents a certain function from the domain τ_1 to the function range τ_2. λ is thus used in records of structured expressions that denote functions (in common mathematical notation, "$x \mapsto ...$" seems to correspond to "λx"). For a somewhat different example, the untyped term "$[\lambda x[x + 1]]$" represents a function whose domain consists of (all) values for the variable x (λ binds all free occurrences of "x" in the 'body' of the λ-abstraction), and whose function range consists of numbers one gets by calculating $x+1$ for each possible value of x; the term "$[\lambda x[x+1]]$" thus represents the function of addition.

ii. With regards to the notational variants of $\mathcal{L}_{\lambda_{Ch}}$, the expression "$x : \tau$" is more frequent than "x^{τ}". Moreover, many writers drop type indications, provided rendering of the type-theoretic correctness of a term is not affected. Curry (e.g. [90]), however, proposed a radically different approach, since he used a notation free from of all type annotations. But types of his terms are implicitly determined by the explicit *type context* Γ, which contains type declarations such as "$x : \tau$". It is known as *typing à Curry* ("Curry-typing", "implicit typing").

Curry thus implemented *type polymorphism*, for e.g. the type-theoretically polymorphous symbol "$=$" obtains its particular type from the surrounding context. Church's notation, on the other hand, has infinitely many "$=^{\tau}$" symbols, each with a distinct type τ. Type polymorphism resembles Russell's (e.g. [353, 352]) idea of *typical ambiguity* as the essential feature of functions.

---

[2] In BNFs, I systematically suppress quotation marks.

iii. Typing thus puts restrictions on the formation of well-formed terms. The restrictions were criticised (e.g. by Quine [314]), but any notation employing e.g. type-polymorphous symbols (and/or reference to VCP) seem to avoid such objections.

2.2.3 Applications of type theories and typed λ-calculus

As indicated above, λ-calculus can serve as a tool for the foundation of mathematics, which is how Church [70, 68, 72] originally conceived it. For example, natural numbers are coded by *Church's numerals*, i.e. terms of the form "$[\lambda f[\lambda x[f^n\, x]]]$" (where n iterations of "f" in the 'body' of the term equals the defined natural number), the truth values "true" and "false" are coded by the terms "$[\lambda x[\lambda y.x]]$" and "$[\lambda x[\lambda y.y]]$", etc.; finally, one represents *recursive functions*. One of the famous results of λ-calculus is the *undecidability of first-order logic* proved by Church [71, 69].

Many comparable results have been achieved within *combinatory logic* which is translatable with λ-calculus, see Schönfinkel [363] (a pioneering article on the topic), Curry [89], Curry and Feys [90]. Recently, parallels between certain STTs and *categories* have been studied (esp. Cartesian closed categories; see e.g. Jacobs [185]); one particularly interesting piece of research is framed within homotopy type theory, see Voevodsky [441]. A remarkable semantics for untyped λ-calculus was developed by Scott (e.g. [368]).

For an introduction to Church's STT, see Benzmüller and Andrews [38]; for an introduction to the contemporary TTs, see Coquand [80], Cardone and Hindley [56] or Turner [435]. For a well-appreciated introduction to λ-calculus and combinatory logic, see Hindley and Seldin [174]; also see the exhaustive books by Andrews [10], Barendregt, Dekker and Statman [24] (typed λ-calculus), and mainly Barendregt [23] (untyped λ-calculus).

It is important to add that Churchian STT only utilises *functional types*, but the contemporary TTs accept other kinds of types. Currently, one of the most popular TTs is *dependent TT* (see e.g. Barendregt, Dekker and Statman [24] or other writings on such *pure type systems*), whose types (as terms) may contain type variables.

λ-calculus is highly important in computer science, esp. as a model of *functional programming* (see Michaelson [254] or Pierce [301]) because untyped λ-calculus corresponds to languages such as LISP or Scheme, while typed λ-calculus corresponds to languages such as Haskell or ML. Computer scientists have investigated many properties of particular λ-calculi.

Another important research area within TT is the *Curry-Howard correspondence* (or *C-H isomorphism*), which is based on i. similarities of type terms such as "$\tau_1 \longrightarrow (\tau_2 \longrightarrow \tau_1)$" and tautologies of (intuitionistic) implicational fragment of propositional logic (called *formulae-as-types correspondence* or *propositions-as-types correspondence*, PAT), and ii. similarities of proofs and programs (*proofs-as-programs correspondence*). See Sørensen and Urzyczyn [384] or Girard, Taylor and Lafont [156] who also explain the important correspondence related to *proof theory*: enabling easier proof search and reductions of proofs (e.g., cut-elimination). For TT in relation to proof theory, see e.g. Nederpelt and Geuvers [273], Kamareddine, Laan and Nederpelt [201].

Within investigations of *intuitionistic foundations of matematics*, the *intuitionistic (constructive) TT* by Martin-Löf [245], which uses the Curry-Howard correspondence, is famous (for applications in computer science, see Nordström, Petersson and Smith [275]). For another approach, see e.g. Fletcher [130] or Girard, Taylor and Lafont [156]. For discussion of RTT from intuitionistic viewpoint see Palmgren [287].

Another important area is *automated theorem proving* where *proof assistants* based on modern TTs (esp. Coq, see Coquand and Huet [81], [399], Nuprl, see [402]) and also on Churchian STT (esp. HOL, see its manual [400], Isabelle, see its manual [401]) are prominent.

As mentioned in 1.2.2,4.2, 4.2.1, TT is also utilised as a tool for the *formalisation of language*, see Montague [258], and even Tichý [419]. For recent works, see e.g. Muskens [270], Fox and Lappin [134], Chatzikyriakidis and Luo [62].

2.2.4 Ramified type theory

Russell's [356] first TT was simple, since it did not split the type of 'propositions' into particular (sub)types. It is thus possible to quantify over the whole type of 'propositions'; as a result, it is easy to construct a paradox misusing it. One such paradox, namely the *Russell-Myhill paradox* (Russell [356], Klement [214]), led Russell to the renunciation of STT.

As a replacement of STT that would be immune to logical paradoxes and suitable for foundations of mathematics, Russell proposed the first *ramified TT* (*RTT*), see Russell [353] and Whitehead and Russell [445] (with some modifications). The essential feature of Russell's RTT is that not only the type of 'propositions', but also the type of monadic 'propositional functions', the type of dyadic 'propositional functions', etc., are each ramified, i.e. split into proper types that differ in their *order*.

Convention 2 (Order, order of type, order of object)

Order is a natural number $k \in \mathbb{N}$ that is assigned to a type or types within a given TT. An object that belongs to such a type of order k has k as its order, too. A type of order k will be called a "*kth-order type*"; an object of order k will be called a "*kth-order object*".

Russell understood 'propositional functions' as intensional entities; he considered them as linguistic entities that contain predicates, variables, connectives and quantifiers. With the exception of individuals, Russell eliminated all extensional objects (incl. sets); the eliminated entities were represented in RTT by certain 'propositional functions'.[3]

The stratification of 'propositional functions' (incl. 'propositions' which can be viewed as their nulary case) prevents the possibility of a 'propositional function' that would quantify over a type of which it is a member. This is a practical implementation of VCP; Russell adopted the idea from Poincaré (see also 2.2.5). For example, 'propositional functions' exclusively involving variables for individuals are members of types of 1st-order functions; such 1st-order types can only be quantified over by 'propositional functions' of an order greater than 1.

Church's [66] reconstruction of Russell's RTT, the *theory of r-types*, presents a convenient replacement of Russell's somewhat convoluted definitions of type and order (see also Laan and Nederpelt [224]):

Definition 12 (Church's RTT)

1_{t_i}. The collection i containing variables for individuals is an *r-type*.

$1_{t_{ii}}$. If $\tau_1, \tau_2, ..., \tau_m$ are r-types, then $(\tau_1\tau_2...\tau_m)/n$ is an *r-type* containing variables for m-ary functions (for $1 \leqslant m$) of level n, for $1 \leqslant n$.

2_{o_i}. The *order* of each variable for individuals equals 0.

$2_{o_{ii}}$. The *order* of r-type $(\tau_1\tau_2...\tau_m)/n$ equals $N + n$, where N is the greatest among the orders of r-types $\tau_1, \tau_2, ..., \tau_m$ (while $N = 0$, if $m = 0$).

3. Nothing is 1. an *r-type*, or 2. an *order of r-type*, unless it so follows from 1_{t_i}. or $1_{t_{ii}}$., or 2_{o_i}. or $2_{o_{ii}}$.

Remarks. To illustrate the definition, a (simple) binary 'propositional function' that contains two individual variables is of r-type $(ii)/1$, while (simple) unary 'propositional function' that contains a variable belonging to the r-type is of r-type $((ii)/1)/1$. 'Propositions' belong to r-types of the form $()/n$.

Russell's RTT was then rejected for various reasons: that it is too complicated, that it treats 'spurious' intensional entities rather than the extensional ones, that 'epistemic paradoxes' (the Liar paradox, Russell-Myhill paradox, etc.) lie outside mathematics, and STT therefore suffices. Another criticism targeted Russell's Axiom of Reducibility (see comments in 2.4.4 below) and the impossibility of reconstructing Dedekind's cuts as well as real numbers within the system of Principia Mathematica.

---

[3] Russell's logic is now intensively studied, see e.g. Landini [225], Linsky [240], or Irvine [184] for an introduction.

Later development attempted to avoid such objections, and new RTTs have been formulated. See Copi [79], Chihara [65], Church [66], Laan and Nederpelt [224], or Thomason [408] where one can also find motivation for RTT. Recent RTTs include the RTT by Kamareddine, Laan and Nederpelt [202, 200], formed on the basis of modern TTs, or Tichý's [422] TTT, in which only one type is ramified, while STT is embedded in it, and so TTT is capable of treating functions in the extensional sense, which prevents most of the criticism raised against Russell's RTT.

For the goals of this book, which are concerned with 'propositions' and belief, STT is therefore insufficient. An adequate individuation of 'propositions', which precludes related paradoxes, can only be offered within RTT. I thus agree with Church that

> If, following early Russell, we hold that the object of assertion or belief is a proposition and then impose on propositions the strong conditions of identity which it requires, while at the same time undertaking to formulate a logic that will suffice for classical mathematics, we therefore find no alternative except for ramified type theory with axioms of reducibility.
>
> Church [75], p. 521

2.2.5 Vicious Circle Principles

Now allow me to briefly discuss VCP, for it is not sufficient to view TTT in this book as a convenient, yet arbitrary logical tool for the analysis of 'propositions': TTT must be understood as well-founded.

Russell [445] first noticed that entities cannot be specified with the help of themselves, since such a specification would be circular, and so the very existence of the specified entity could then be questioned. The corresponding Ur-principle governing the formation (specification) of entities is thus hardly open to doubt.

> *The Principle of Specification.* No entity can sufficiently be specified if one specifies it using the entity just specified.

Remark. To illustrate, the specification of a function f requires the determination of its arguments and values (and the links between them); however, f's specification would be impossible, if f would be among the arguments or values, because f has not yet been specified.[4]

[4] Russell [445] noted that a function which consists of an infinity of arguments/values cannot be specified by the enumeration of its arguments/values, but using a condition (i.e. an 'intensional' specification). Moreover, a specification cannot deploy an entity whose specification presupposes the entity that has been just specified (my formulations of VCPs should be supplemented in this sense). Russell's reasons for ramification are discussed e.g. in Goldfarb [158], Hodes [180].

The Principle of Specification clearly entails VCP in all its variants. Since TTT involves both extensional (e.g., functions) and intensional entities (namely constructions; some of them resemble Russell's 'propositional functions'), it is possible to recognise four variants of VCP in TTT. Unlike Russell's versions of VCP [445], my variants of VCP differ in the category of entities they consider.[5]

i. *The Functional VCP.* No function can be its own argument or value (or their parts).

ii. *The Constructional VCP.* No construction can (on any valuation) construct itself or a construction which contains it.

Remarks. The reason for the adoption of the Functional VCP was suggested in the previous remark. It is obvious that this VCP justifies the hierarchy of functions in STTs. The Constructional VCP justifies the hierarchy of constructions. To illustrate, consider e.g. the variable c, which is a construction (see 2.3.1, 2.6) that constructs some constructions. For the specification of c, all its possible values must be specified, and so on no valuation may c construct itself, or a construction constructing it, or containing it.

VCPs i. and ii. imply two others VCPs that are also implemented in TTT:

iii. *The Functional-constructional VCP.* No construction can (on any valuation) construct a function of which it is an argument or value.

iv. *The Constructional-functional VCP.* No function can involve among its arguments or values a construction that constructs the function.

2.3 The language of constructional pre-terms $\mathcal{L}_*$

As mentioned above, each TT imposes restrictions on the formation of terms, which brings about complexity in the resulting formal language. To proceed gradually, one starts with a simpler language that consists of so-called *pre-terms* ("raw terms"). Pre-terms are then 'decorated' by type terms and those terms which match the type prescriptions of a given TT are selected for the resulting language.

[5] Gödel's [157] objections against Russell's variants of VCP, see also e.g. Jung [198], do not seem to be applicable to my VCPs. Moreover, Ramsey's [341] notorious objection, according to which there is nothing vicious on the definition of the tallest man by quantifying over all men, is misguided as a criticism of VCP (Tichý [422]), since men are mere individuals, not functions as 'intensional entities' which were discussed by Russell.

Definition 13 (Language of constructional pre-terms, $\mathcal{L}_*$)

$C_{(i)} ::= x \mid {}^0X \mid {}^1X \mid {}^2X \mid [CC_1...C_m] \mid [\lambda x_1...x_m.C]$

where "X" can be a certain "$C_{(i)}$".

Convention 3 (Names of kinds of constructions)

Term	*name of a kind of the represented construction*
"x"	*variable*
"${}^0X$"	0-*execution*
"${}^1X$"	1-*execution*
"${}^2X$"	2-*execution*
"$[CC_1...C_m]$"	*composition*
"$[\lambda x_1...x_m.C]$"	*closure*

Remark. Tichý [422] introduced 0-execution under the name "*trivialization*" and ${}^1X$ and ${}^2X$ under the names "*execution*" and "*double execution*", respectively. Hierarchy of executions, whose bottom (0-ary) case is 0-execution, was studied in Raclavský [321], see also Pezlar [296].

Notational convention 4 (Infix notation of binary operators)

Symbols for familiar binary logical and mathematical operators will be written in the infix manner.

Remark. "$\wedge$", "$\rightarrow$", "$+$", "$=$", "$\leqslant$", "$\cong$", "$+$", "$-$" and "$\div$" are examples of such operators, and so I will write e.g. "$[[{}^0 1\, {}^0{+}\, {}^0 2]\, {}^0{\leqslant}\, {}^0 3]$" in lieu of "$[{}^0{\leqslant}\ [{}^0{+}\, {}^0 1\, {}^0 2]\, {}^0 3]$".

Notational convention 5 (Suppressing brackets of nested closures)

Brackets of nested closures will often be supressed.

Comments.

i. $\mathcal{L}_*$ obviously differs from the language of pure λ-calculus in its having the special terms "${}^0X$", "${}^1X$", "${}^2X$", whose semantics will be described in 2.6, 2.6.3. Further, $\mathcal{L}_*$ allows strings "$C_1...C_m$" (in which "$C_{(i)}$" can be "x") because m-ary functions are adopted in TTT (2.4.3).

ii. Constructions form the intended fine-grained, intensional *semantics* for $\mathcal{L}_*$ and languages based on it. The objects constructed by the constructions are not the primary semantic values for terms of $\mathcal{L}_*$, they only form the *denotational semantics* for $\mathcal{L}_*$ (or $\mathcal{L}_{\mathsf{TTT}}$); see 2.6.

iii. An important difference of Tichý's 'logic of constructions' from λ-calculus consists in that, in λ-calculus, every occurrence of a variable in a term counts as a part of the term. For example, the term

"$[\lambda x.xx]$"

contains 3 occurrences of the variable "x", which are all bound by the operator λ (which is, along with brackets and ".", a syncategorematic expression). As a term of $\mathcal{L}_*$ (or $\mathcal{L}_{\mathsf{TTT}}$), $[\lambda x.xx]$ is a record of a construction that only contains 2 occurrences of a variable, because the *binding sequence* "$\lambda x.$" is not a part of the whole construction, but of the term. The whole construction contains variables as (sub)constructions; the variables thus cannot be mere letters (see the next section).

2.3.1 Valuation and v-constructing

To treat variables not as linguistic expressions, but as genuine constructions, Tichý [422] had to propose a novel definition of *valuation*. He generalised Tarski's [398] idea according to which a valuation is a sequence of objects that can be assigned to variables.

Definition 14 (Valuation, v)

Let $\tau_1, \tau_2, \tau_3, \ldots$ be distinct types. For every type τ_i there exists n possible infinite (so-called) τ_i-sequences of objects of type τ_i. Every *valuation v* is a matrix consisting of (an infinity of) τ_i-sequences, while for every type τ_i, exactly one τ_i-sequence becomes a member of v.

Remarks. The two valuations v_j and v_k differ if there is a difference in at least one of their τ_i-sequences. Here is an example of a (schematic) valuation v (let n be the number of types):

τ_1-sequence:	$X_1^{\tau_1}, X_2^{\tau_1}, X_3^{\tau_1}, \ldots, X_m^{\tau_1}, \ldots$
τ_2-sequence:	$X_1^{\tau_2}, X_2^{\tau_2}, X_3^{\tau_2}, \ldots, X_m^{\tau_2}, \ldots$
τ_3-sequence:	$X_1^{\tau_3}, X_2^{\tau_3}, X_3^{\tau_3}, \ldots, X_m^{\tau_3}, \ldots$
$\vdots$	$\vdots$
τ_n-sequence:	$X_1^{\tau_n}, X_2^{\tau_n}, X_3^{\tau_n}, \ldots, X_m^{\tau_n}, \ldots$

Convention 6 (Assignment of an object by valuation, '$v(X/x)$')

"$v(X/x)$" is short for "the valuation v that assigns an object X to the variable x"; this can be extended to "$v(X_1/x_1, X_2/x_2, \ldots, X_n/x_n)$".

For semantics of $\mathcal{L}_{\mathsf{TTT}}$ (2.6) I will use the following notion which should not be confused with that of τ-sequence.

Definition 15 (Valuation sequence, sq^τ)

Valuation sequence sq^τ is a function that maps all variables v-constructing objects of type τ to members of τ-sequence.

2.3.2 v-constructing, v-congruence, v-improper constructions

With the notion of valuation at hand, some further notions related to constructions can be offered (adopted from Tichý [422]).

Constructions may contain variables (as constructions), and so their constructing depends on valuation. The idea generalises to include the case of every construction.

Convention 7 (v-**Constructing**)

Every construction C constructs in dependence on a certain valuation v, I will briefly say that "C *v-constructs*".

Remark. Since the notion of construction and v-constructing coincide (1.4.1), we could properly speak about *v-constructions.*

Some constructions,[6] e.g. $[3 \div 0]$, can fail in v-constructing in the sense that they v-construct nothing at all. Some constructions, e.g. $[n \div 0]$, are abortive even on every v.

Definition 16 (v-**Improper construction**)

A construction C is *v-improper* iff C v-constructs nothing at all. Otherwise, C is *v-proper.*

Some constructions, e.g. $[1 + 2]$, $[[4 - 2] + 1]$ and $[\sqrt{}9]$, v-construct the same object, or nothing at all, e.g. $[x \div 0]$ and $[[x + y] \div 0]$.

Definition 17 (v-**Congruence**, $\cong$)

Two constructions C_1 and C_2 are *v-congruent* ($\cong$) iff both C_1 and C_2 v-construct the same object, or they are both v-improper.

Remarks. A full symbol of v-congruence should be "$\underset{v}{\cong}$", but for the reasons of brevity I suppress "v", since it is often implicitly given in context (similarly, I omit an indication of orders of related constructions).

2.4 The main part of Tichý's type theory

2.4.1 Type base

Every type hierarchy of entities must treat some objects as logically primitive (regardless of their real complexity), and thus as members of their type *base.*

6 In this section, I utilise an agreement leading to suppression of "^{0}", see 2.5.

Definition 18 ((Type) base, B)

(Type) base B is any set of pairwise disjoint sets that contain logically primitive objects of a certain 'kind'.

Remark. To illustrate, $B = \{\{I_1, I_2, I_3\}, \{\mathtt{T}, \mathtt{F}\}\}$ is a base that contains a certain set of individuals and the set of the truth values.

2.4.2 The language of typed terms $\mathcal{L}_\tau$

Here is the second language, $\mathcal{L}_\tau$, I will utilise. The semantics for it is provided by the definition of TTT (2.4.4).

Definition 19 (Language of typed terms, $\mathcal{L}_\tau$)
$\tau ::= \tau_B \mid *_k \mid (\tau_0\tau_1\tau_2...\tau_m)$

Remarks. Since in TTT m-ary functions are not reduced to unary ones as in STT (2.4.3), "$(\tau_0\tau_1\tau_2...\tau_m)$" corresponds to the more spacious expression $\langle\tau_1, \ldots, \tau_m\rangle \mapsto \tau_0$. $\mathcal{L}_\tau$ does not have type variables, as is common in many contemporary TTs. Two warnings concerning notation: "$*_{(i)}$" neither denotes the type of types, as in modern TTs ("pure type systems", e.g. Barendregt, Dekker and Statman [24]), nor ramified types of Kamareddine, Laan and Nederpelt [201].

2.4.3 Partial functions

Unlike many STTs, TTT treats both *total* and *partial functions*, in the frequent meaning of the term "partial function":

Definition 20 (Partial function)

A *partial function* is a function that is undefined at least for one its argument, i.e. it maps it to no value.

It is important to observe that TTT implements partiality on two levels:

i. it treats partial functions

ii. it treats v-improper constructions

Sometimes there is a dependence of ii. on i., since some constructions (e.g. the construction $[3 \div 0]$) are v-improper because a certain function v-constructed by some its subconstruction is partial.

Writers who work with partiality mention several reasons for its adoption in foundations of mathematics and computer science (see e.g. Feferman [123]): a. the claim that the value of $3 \div 0$ is an infinite value, or even an arbitrary value, lacks substantiation – it is thus more acceptable to admit that $3 \div 0$ yields no value; b. seeking a database does not bring a result ('no matches found'); c. a program gets stuck (unexpected freezing, cycling). d. Within the analysis of natural language, Strawson's [393] theory of gappy (truth-valueless) sentences such "The King of France is bald." is often mentioned (see 4.5.5 for discussion). I thus agree with Tichý's view that

> Partial type theory seems to provide a medium which yields an analysis for any linguistic expression, and affords a universal explication of logical entailment.
>
> Tichý [415], p. 59

Various writers implemented partiality in their logical systems, e.g. Tichý [415, 416], Blamey [44], Moggi [255], Farmer [121], Asher [16], Lepage [233], Lappiere [227], Muskens [270], Langholm [226], Thissje [404], Duží [103], Areces, Blackburn, Huertas and Manzano [11], Běhounek and Daňková [54], Novák [276].

I add that *Schönfinkel's reduction* (Schönfinkel [363]), known as *currying* after Curry, of m-ary (mulitargument) functions to corresponding unary functions is usually adopted in STT, which is why they only manipulate unary functions. However, Tichý [415] proved that the reduction fails if partial functions are admitted.[7] He therefore employed *m-ary functions* both in his STT [415] and TTT [422]. I will call arguments of m-ary functions "*(m-ary) strings*" to avoid confusion with m-tuples which I will identify with certain partial functions (2.5.1).

2.4.4 Definition of TTT

The following *Definition*$_{\mathsf{TTT}}$ (as I will call it) is Tichý's [422] Definition 16.1 with some unsubstantial changes. There, Tichý elegantly defined three terms: "type", "order of construction", "order of type".

As I illustrate in comment i. below, Tichý's formulation can be extended so that the languages $\mathcal{L}_\tau$ and $\mathcal{L}_*$ are explicitly referred to (Tichý did not define such languages). The adjustment only makes it more obvious that Definition$_{\mathsf{TTT}}$ provides the semantics for $\mathcal{L}_\tau$ because it assigns certain sets to the type terms of $\mathcal{L}_\tau$ as semantic values. The result of such an adjustment (described in comment i. below) would lead to an unnecessary intricacy of the definition, which is why I prefer to keep its original version.

[7] In proving the invalidity of Schönfinkel's reduction, Tichý [415] refers to the existence of a binary partial function f which maps binary strings (couples) of truth values to truth values. The value of $f(x, y)$ is y, if $x = \mathtt{F}$ (False); f is undefined otherwise. The reduction of f to a unary function is not unique, since two unary functions correspond to f. Each of them maps $\mathtt{F}$ to the identity function, but one of the functions is undefined for $\mathtt{T}$ (True), whereas the second function maps $\mathtt{T}$ to the truth function which is undefined for all its arguments.

Here is the scheme of the definition:

1. First, types of 1st-order objects are defined (this is in fact Tichý's [415] STT).
2. It is then specified which constructions belong to which type of constructions of a certain order (this way, constructions are related to types of objects they v-construct).
3. Finally, in the inductive step, types of orders greater than 1 are defined, utilising also the types of constructions.

Definition 21 (Definition$_{\mathsf{TTT}}$)

Let B be a base.

1. (*1st-order types*)

$(t_1\mathrm{i})$	Every member of the base B is a *type of order* 1 *over* B.
$(t_1\mathrm{ii})$	If $0 < m$ a $\tau_0, \tau_1, \ldots, \tau_m$ are types of order 1 over B, the set $(\tau_0\tau_1 \ldots \tau_m)$ of all m-ary total and partial functions from $\tau_1, \ldots, \tau_m$ to τ_0 is also a *type of order* 1 *over* B.
$(t_1\mathrm{iii})$	Nothing is a *type of order* 1 *over* B, unless it so follows from $(t_1\mathrm{i})$ or $(t_1\mathrm{ii})$.

Let τ be any type of order n over B.

2. (*nth-order construction*)

$(c_n\mathrm{i})$	Every variable x ranging over τ is a *construction of order* n *over* B. If X is a member of type τ, then ${}^0X$, ${}^1X$ and ${}^2X$ are also *constructions of order* n *over* B.
$(c_n\mathrm{ii})$	If $0 < m$ a $C_0, C_1, \ldots, C_m$ are constructions of order n, then $[C_0C_1 \ldots C_m]$ is a *construction of order* n *over* B. If $0 < m$, and C and also variables $x_1, \ldots, x_m$ are constructions of order n over B, then $[\lambda x_1 \ldots x_m.C]$ is also a *construction of order* n *over* B.
$(c_n\mathrm{iii})$	Nothing is a *construction of order* n *over* B, unless it so follows from $(c_n\mathrm{i})$ or $(c_n\mathrm{ii})$.

Let $*_n$ be a set of constructions of order n over B.

3. (($n+1$)*st-order types*)

(t_{n+1}i)	The type $*_n$ is a *type of order* $n+1$ *over* B. Every type of order n is also a *type of order* $n+1$ *over* B.
(t_{n+1}ii)	If $0 < m$ and $\tau, \tau_1, \ldots, \tau_m$ are types of order $n+1$ over B, then the set $(\tau\tau_1 \ldots \tau_m)$ of all m-ary total and partial functions from $\tau_1, \ldots, \tau_m$ to τ is also a *type of order* $n+1$ *over* B.
(t_{n+1}iii)	Nothing is a *type of order* $n+1$ *over* B, unless it so follows from (t_{n+1}i) or (t_{n+1}ii).

Convention 8 (kth-order construction, kth-order object)

A construction of order k (i.e. a member of $*_k$) is called in brief a "k*th-order construction*". An object belonging to a type of order k is called in brief a "k*th-order object*".

Comments.

i. As mentioned above, points 1. and 3. in fact present the *semantics for* $\mathcal{L}_\mathcal{T}$. To make its formulation precise, one should deploy phrases such as: a. the term "τ_B" denotes such-and-such a set $S_O \in B_{\mathsf{TTT}}$ of objects; b. the term "$*_k$" denotes such-and-such a set S_C of constructions over B_{TTT}; c. the term "$(\tau_0\tau_1, \tau_2, ..., \tau_m)$" denotes such-and-such a set S_F of m-ary functions over B_{TTT}.

ii. The notion of order of constructions is dependent on the notion of type, i.e. a type of a certain order: C is a kth-order construction, if it v-constructs a kth-order object. Thus, speaking about orders of constructions simplifies the way of speaking about their types.[8]

iii. Every construction of a certain member of the hierarchy of TTT is also a member of that hierarchy. It holds that a construction C, which v-constructs a certain kth-order object, is of a (strictly) greater order than k, namely $k+1$ (as is specified in point 2., every kth-order construction C is a construction of a certain $(k-1)$st-order object (if any)).

[8] Here TTT provides an apt solution to the problem recently raised by Whittle [447] (see also Sbardolini [361]): if logical operators are 'propositional functions', no ordinary 'propositions' involving them are possible in RTT, for hierarchisation of logical operators as 'propositional functions' into orders prevents that. For example, the RTT transcription of the familiar formula $\neg\neg\varphi$ would contain two negations, one of order 2 and one of order 1. In TTT, $\neg$ or $\wedge$ are not 'propositional functions', but functions (as mappings) of types (oo) or (ooo), i.e. certain 1st-order objects; their 0-executions can be concatenated with various other constructions of 1st-order objects without any obstacles.

iv. We may perhaps say that $\text{Definition}_{\mathsf{TTT}}$ provides *typing* objects (incl. constructions). According to usual praxis in contemporary TTs, however, only terms are typed: the expression "$C : \tau$" means that the term "C" is assigned the type term "τ". In TTT, such a typing of constructions (or terms representing them) would not be informative enough, because it would always be of the form "$C : *_k$". Moreover, the fact so recorded, namely that C is of type $*_k$, can be derived from a typically much more useful piece of information about C, namely from the fact that C v-constructs an object (if any) of a kth-order type τ.[9]

In the communication of such a piece of information, I will take into account the fact that C need not to v-construct any object of a particular type τ, but that C hypothetically may v-construct one. For example, the construction written as "$[3 \div 0]$" does not v-construct anything (and 'real nothing' does not conceivably belong to any type), but at least hypothetically, it should be a number.

Convention 9 ('C/τ', 'C is a τ-construction')

That C v-constructs, or that C may at least hypothetically v-construct, an object of type τ, is briefly written as "$C/^v\,\tau$". In this record, "^{v}" will be suppressed, provided a precise determination of a particular v is not needed in the context. Furthermore, "$C_1, C_2/\tau$" is short for "C_1/τ; C_2/τ". "C/τ" is occasionally expressed by "C is a τ-*construction*".

Further complications with type assignement will be discussed in section 2.5. Here I close the topic with the remark that $\mathcal{L}_{\mathsf{TTT}}$'s terms can be paired with full type annotations such as "$C/*_k : \tau$" (for another method, see Pezlar [295]), but I do not employ such a proposal in this book.

v. TTT has a number of interesting logical properties:[10]

(1) TTT involves STT, though it is ramified.

(2) TTT is ramified simply. As a result of this, type $*_k$ includes all kth-order constructions regardless of the various types of objects they v-construct. This enables quantification over constructions of $*_k$ without the restrictions so typical of various RTTs. For comparison, Russell's RTT disallows quantification over all 'propositional functions' of a certain order k because kth-order 'propositional functions' of individuals and kth-order 'propositions' belong to pairwise disjoint types.

---

9 To illustrate the point, it follows from $\text{Definition}_{\mathsf{TTT}}$ that a certain type is a 1st-order one if it satisfies point 1. For example, the type $(\iota\iota)$, where ι is a member of B, is a 1st-order type. Points 2 and 3 entail that types of orders greater than 1 have their particular order k indicated in the record of the type $*_k$ which is that 'part' of the type in question which has the greatest order in it: e.g., the type $(*_2\iota*_1)$ is a 3rd-order one because its greatest "k" in its record equals 2.

10 Raclavský [328], Raclavský, Kuchyňka and Pezlar [339].

(3) Definition$_{\mathsf{TTT}}$ enables instances of TTT that differ in which objects over which base they harbour. For example, B of a certain instance may contain $\mathbb{N}$ (not $\mathbb{N}^0$), so the division function involved in such an instance of TTT is $\div_{\mathbb{N}}$, which is not a partial function (unlike $\div_{\mathbb{N}^0}$). In addition to individuals and truth values, some instances of TTT may contain a base type of possible worlds (or time instants, etc.); such instances are thus suitable for various (higher-order) modal logics.

(4) Definition$_{\mathsf{TTT}}$ also entails that types are *cumulated* in orders in an upward direction, which entails that every kth-order object is also a $(k+1)$st-order one. As a result of point 2., the following principle holds in TTT.

Fact 1 (The Principle of Cumulativity of Constructions, PCC)

Every kth-order construction is also a $(k+1)$st-order one.

This cumulativity resembles the cumulativity which is explicitly present in Church's RTT [66]. It allows that type $*_k$ also contains all the constructions of orders equal to, or lower than $k-1$ (otherwise, the system would be unduly restrictive), i.e., $*_1 \subset *_2 \subset ... \subset *_n$, which means that types for constructions are not pairwise disjoint. When speaking about orders of constructions, I will usually mean their lowest, *native order*, though C can be a part of another construction in which C occurs with its greater order.

(5) A specific topic is *reducibility* that Russell ([353], [445] with Whitehead) expressed via his *Axiom of Reducibility* (see e.g. Goldfarb [158] for more). He himself considered it nonevident, and, after suspecting that RTT with the axiom collapses to STT (see Ramsey [341]) and further problems, he and Whitehead rejected it [446].[11] Within TTT, there hold various facts that resemble the Axiom of Reducibility, but some of them are much stronger, and so I will prefer to speak about *general reducibility*:[12]

[11] Recall that Russell individuated properties as 'propositional functions', so he could not adequately formalise the sentence "Napoleon has all the properties of great generals.", for it presupposes the possibility of quantification over all properties. Such a quantification is impossible, for 'propositional functions' are split into various types, i.e. in the ranges of type-theoretically different variables. But the Axiom of Reducibility says that, when quantifying over properties, one can limit oneself to the order 1, since the respective type contains equivalent mates of the higher-order properties. Let me add that in THL 4.4.1, properties of individuals are identified with certain functions, i.e. objects of the order 1, the sentence in question can be thus formalised without any obstacles and without a use of the Axiom of Reducibility.

[12] The formulation speaks about a kth-order object, which is, however, also of order m, since it is v-constructed by an $(m+1)$st-order construction; but I allude to the fact that it is of (possibly) lower order k while it is v-constructed by a kth-order construction. The formulation is intended to cover also v-improper constructions (to detect k and m of such constructions one should investigate orders of its subconstructions and context of their

Fact 2 (The General Principle of Reducibility, GPR)

For every $(k+j)$th-order construction C^{k+j} (for $1 \leqslant k$, $1 \leqslant j$) of a certain kth-order object X, there exists a v-congruent kth-order construction C^k of the same object X.

For proof of the GPR (Raclavský [328, 339]), it is enough to become conscious of the facts that i. TTT is not limited to any definite vocabulary and $\mathcal{L}_{\mathsf{TTT}}$ is a very comprehensive language form, and ii. there is a vast amount of constructions, and so it is always possible to find enough constructions satisfying the GPR. For this, TTT's capability to prove v-congruence of constructions is crucial. In TTT, we may thus state the identity of denotational values of terms, which cannot be achieved within Russell's RTT.

Note also that the GPR is more general than Russell's Axiom of Reducibility; especially, it is not limited to (im)predicative constructions. Unlike Russell's axiom, which can be adopted or discharged from his RTT, the GPR expresses an inherent property of TTT. Note also that the Axiom of Reducibility was introduced to bypass type restrictions and so enable quantification over numbers and properties as 'propositional functions'; given the explication of (real) numbers and properties within TTT (namely within THL), TTT has no such need.

(6) Finally, within any instance of TTT one can find various *impredicative constructions* (according to various definitions of the notion, see e.g. Feferman [124]; see also Hazen [168] for predicative logics).

2.5 The language $\mathcal{L}_{\mathsf{TTT}}$

After resolving the issue with the v-improper constructions discussed in the previous subsection in point iv., we could identify $\mathcal{L}_{\mathsf{TTT}}$ with the following language, which emerges from $\mathcal{L}_*$ via the 'decoration' of its terms using $\mathcal{L}_\tau$'s terms:

Definition 22 (Language of typed constructional terms, $\mathcal{L}_{*_\tau}$)

$$C^\tau ::= x^\tau \mid ({}^0X)^\tau \mid ({}^1X)^\tau \mid ({}^2X)^\tau \mid [C_0^{\tau_0} C_1^{\tau_1} ... C_m^{\tau_m}]^\tau \mid [\lambda x_1^{\tau_1} ... x_m^{\tau_m}.C_0^{\tau_0}]^\tau$$

where "$\tau_{(i)}$" conforms to Definition$_{\mathsf{TTT}}$.

For this and subsequent languages I introduce the following convention:

use, which would ramify the above formulation of the GPR).

Notational convention 10 (Suppression of the 0-execution symbol)

If it is unambiguous that X of ${}^0X$ is a 1st-order object, the symbol "^{0}" will be suppressed, provided ${}^0X$ occurs within the context of another construction; but I will write "${}^{(0)}X$", provided ${}^0X$ stands outside any construction. If it does not provide misunderstanding, I extend the convention also for cases when X in ${}^0X$ is a kth-order object, for $1 < k$. Bold face often indicates the suppression of "^{0}".

Remarks. For example, "${}^0[{}^03\,{}^0\div\,{}^00]$" and "^{03}" will be written in brief as "${}^0[3\div 0]$" and "${}^{(0)}3$", respectively. In the record of a solitude-standing construction ${}^{(0)}X$, one cannot suppress "^{0}" because it cannot be decided whether "X" stands for the construction ${}^0X$, or for the object X. In the definitions below, ${}^0X$ sometimes occurs outside the context of any construction, but since each such definition is general, X need not be a 1st-order object, which is why ${}^0X$'s record cannot be adjusted to "${}^{(0)}X$". Examples of kth-order objects whose 0-executions are written without "^{0}" (if they occur within records of other constructions) are Bel^k and K^k, so I write e.g. "$[\mathbf{Bel}^k\, x\, c^k]_{tw}$" instead of "$[{}^0\mathrm{Bel}^k\, x\, c^k]_{tw}$".

Comments.

i. In terms of $\mathcal{L}_{*_\tau}$, the symbol "$^\tau$" presents the information otherwise presented using "C/τ". As in modern TTs, type τ can usually be inferred by type inference, provided the types of variables and 0-executions are stated in (so-called) context.

ii. That τ is usually derivable suggests the idea of 'erasing' type indications altogether by means of the *erasing function* ("strip-off function"), defined as

Definition 23 (Erasing function, $|$"C^τ"$|$)

$|$ "C^τ" $| =_{df}$ "C" in which all subterms are results of application of $|$"C^τ"$|$

The result of a systematic application of $|$"C^τ"$|$ to $\mathcal{L}_{*_\tau}$ is the implicitly typed language $\mathcal{L}_{*_{|\tau|}}$, which is not, however, identical to the wholly untyped $\mathcal{L}_*$ (for Definition$_{\mathsf{TTT}}$ is not applied to $\mathcal{L}_*$).

iii. $\mathcal{L}_{\mathsf{TTT}}$ could perhaps be identified with $\mathcal{L}_{*_{|\tau|}}$ (which is discussed in ii.). In this book, I will rather follow Tichý [422], who seems to identify $\mathcal{L}_{\mathsf{TTT}}$ with $\mathcal{L}_{*_{|\tau|}}$ that is enriched mainly by terms of form

$$[C_0^{\tau_0} C_1^{\tau_1} ... C_m^{\tau_m}]^{\tau ?},$$

where

"τ?" indicates the fact that a resulting type τ is not derivable, i.e., that the term is *not typeable* (Raclavský, Kuchyňka and Pezlar [339]). For example, "$[[\lambda x.xx][\lambda x.xx]]$" (a term written down after an application of the erasing function), is not typeable.

An inspiration for the adoption of such terms and the constructions represented by them can be found in Tichý's [422] specification of compositions in which he did not require them to be '*type-theoretically compatible*' (Raclavský [328]). Since Tichý did not comment on the idea, one may considered it as an unintended typo caused by his explanation of types only after the explanation of constructions (we have seen that the reverse order of explaining things is more adequate).

I will also accept terms of forms "$[\lambda x_1^{\tau_1}...x_m^{\tau_m}.C_0^{\tau_0}]^{\tau?}$", "$({}^1X)^{\tau?}$" and "$({}^2X)^{\tau?}$" to increase the expressive power of $\mathcal{L}_{\mathsf{TTT}}$ in comparison to $\mathcal{L}_{*_{|\tau|}}$.[13]

Definition 24 (Language $\mathcal{L}_{\mathsf{TTT}}$)

$$C^{\tau(?)} ::= x^{\tau} \mid ({}^0X)^{\tau} \mid ({}^1X)^{\tau(?)} \mid ({}^2X)^{\tau(?)} \mid [C_0^{\tau_0} C_1^{\tau_1}...C_m^{\tau_m}]^{\tau(?)} \mid [\lambda x_1^{\tau_1}...x_m^{\tau_m}.C_0^{\tau_0}]^{\tau(?)}$$

where "$\tau_{(i)}$" conforms to Definition$_{\mathsf{TTT}}$.

Remarks. Instead of "x^{τ}", there should be "V^{τ}", but "x^{τ}" is more common. Terms of $\mathcal{L}_{\mathsf{TTT}}$ will be written after their simplification by the erasing function |"$C_{(i)}^{\tau(?)}$"|. The semantics of $\mathcal{L}_{\mathsf{TTT}}$ is given in 2.6.

iv. I will often use some terms of $\mathcal{L}_{\mathsf{TTT}}$ extended with the help of *auxiliary annotations*, typically

"X^A" and "x^A",

where "A" codes a certain piece of additional information. For example, it is useful to write "${}^{(0)}\exists^{\tau}$", where "τ" indicates the type of objects over which one 'quantifies'. Or, it is useful to indicate the order k of constructions of type $*_k$ in the record of the variable ranging over $*_k$, e.g. "c^k".

2.5.1 Examples of TTT entities

To illustrate Definition$_{\mathsf{TTT}}$, I am going to offer examples of objects and their types.[14] The arithmetic of natural numbers will be understood as a system of objects over $B_{Ar} = \{\nu, o\}$, where ν is in fact $\mathbb{N}^0$ and $o = \{\mathtt{T}, \mathtt{F}\}$.

First, here are examples of 1st-order types over B_{Ar}:

[13] For example, it possible to code *recursive operators* (e.g., the combinator Y is the construction $[\lambda h[[\lambda x \lambda h[xx]]\ [\lambda x \lambda h[xx]]]]$), which is not possible in strictly typed λ-calculus, which is therefore not *Turing-complete*. Another, well-know solution to the problem was proposed by Plotkin [302].

[14] It extends Tichý's [422] set of convenient examples.

Example 8 (1st-order types over B_{Ar})

$\nu; o; (\nu\nu); (\nu\nu\nu); (o\nu); (oo); (ooo); (o\nu\nu); (o(o\nu))$

If not indicated otherwise, the following examples of functions over B_{Ar} are total functions. For explication of set-theoretic objects which I mention in descriptions see 2.5.2.

Example 9 (Objects of 1st-order types over B_{Ar})

Entity	*type*	*description*
3	ν	number 3
Succ	$(\nu\nu)$	successor function
$\div$	$(\nu\nu\nu)$	(partial) binary function of division
Pr	$(o\nu)$	set of primes (i.e. a certain characteristic function)
$\neg$	(oo)	unary truth function of *negation*
$\rightarrow$	(ooo)	binary truth function of (material) *implication*
$=^{\nu}$	$(o\nu\nu)$	*identity*, i.e. the familiar binary relation between ν-objects
$\forall^{\nu}$	$(o(o\nu))$	*universal quantifier* as a singleton that contains the universal set of numbers
$\exists^{\nu}$	$(o(o\nu))$	*existential quantifier* as the set of all non-empty (total or partial) sets of numbers
$\imath^{\nu}$	$(\nu(o\nu))$	*singularisation function*, i.e. the partial function which maps all one-membered sets to their sole members; it is undefined for all other sets of numbers

Remark. In this book, I will frequently use type-theoretic variants of functions $=^{\nu}, \forall^{\nu}, \exists^{\nu}, \imath^{\nu}$; e.g., $=^{o}$ (the identity between truth values, which could alternatively be denoted by "$\leftrightarrow$"). Further binary truth functions used in this book are $\wedge$ (*conjuntion*), $\vee$ (*disjunction*).

Now let n and s be variables ranging over types ν and $(o\nu)$, respectively.[15]

Example 10 (1st-order constructions over B_{Ar})

${}^{(0)}3; n; s; [\mathrm{Succ}\, n]; [s\, [\mathrm{Succ}\, n]]; [\forall^{\nu} [\lambda n [\mathrm{Pr}\, n]]]; [3 \div n]$

The constructions are of type $*_1$, which is one of 2nd-order types.

Example 11 (2nd-order types over B_{Ar})

$*_1; (*_1\nu); (*_1*_1); (o *_1 *_1); (o(o*_1))$

---

15 The following constructions $v(0/n, \mathrm{Pr}/s)$-construct the following 1st-order objects (in an appropriate order): 3; 0; Pr; 1; F; F; the last construction is $v(0/n, \mathrm{Pr}/s)$-improper.

Example 12 (Objects of 2nd-order types over B_{Ar})

Entity	*type*	*description*
${}^{(0)}3$	$*_1$	0-execution of the number 3
$\mathrm{Triv}^{(*_1\nu)}$	$(*_1\nu)$	*trivialisation function* – also denoted by "$\mathrm{T{\scriptsize RIV}}^{(*_1\nu)}$" – which maps each member of ν to its 1st-order 0-execution (i.e., it maps 0 to ${}^{(0)}0$, 1 to ${}^{(0)}1$, and so on)[16]
Id^{*_1}	$(*_1*_1)$	(unary) identity function operating on 1st-order constructions
$\cong^{*_1}$	$(o *_1 *_1)$	relation of v-congruence between 1st-order constructions

Let $c^1/*_1$ v-construct ${}^{(0)}3$:[17]

Example 13 (2nd-order constructions over B_{Ar})

${}^{00}3; c^1; {}^{(0)}\mathrm{Triv}^{(*_1\nu)}; [\lambda c^1[\mathrm{Id}^{*_1}\, c^1]]$

Example 14 (Objects of 3rd-order types over B_{Ar})

Entity	*type*	*description*
${}^{00}3$	$*_2$	0-execution of the 2nd-order object ${}^{(0)}3$
c^1	$*_2$	variable for 1st-order constructions
$\mathrm{Triv}^{(*_2*_1)}$	$(*_2*_1)$	trivialisation function that maps 1st-order constructions to their 2nd-order 0-executions

2.5.2 Explication of main set-theoretic entities

A worry that TT is not capable of representing set-theoretic entities occurs sometimes in the literature. Yet, those instances of TTT whose base contains types of individuals and truth values are rich enough to represent entities needed for their representation. For example,

Definition 25 (Main set-theoretic entities)

A *set* of objects of the same type is (identified with) a total characteristic function of the objects. A *relation* among m objects is a total characteristic function from m-ary strings ('m-tuples') of objects. A *sequence* of objects is a partial function from natural functions to the objects in question. A *set* containing m objects of distinct types is a total set of total relations which each assign T (True) only to the m-ary string of the objects, while it assigns F (False) otherwise; the relations differ in the order of objects in

[16] The function $\mathrm{Triv}^{(*_1\nu)}$ is Tichý's [422].

[17] The first two constructions $v({}^{(0)}3/c^1)$-construct ${}^{(0)}3$, the latter ones v-construct $\mathrm{Triv}^{(*_1\nu)}$ and Id^{*_1}, respectively.

the strings. An *(ordered) m-tuple* of objects is an m-ary partial function which is defined exclusively for the $(m-1)$-ary string of objects of that m-tuple while the function value is the mth member of the m-tuple.

Remarks. The explication of sets is known from Church's STT, yet another explication of m-tuples is current. Occasionally, I speak about *partial sets*, by which I mean partial characteristic functions.

2.6 Semantics for $\mathcal{L}_{\mathsf{TTT}}$

Now I am going to provide the intended *model-theoretic semantics* for $\mathcal{L}_{\mathsf{TTT}}$. It enables the determination (though not always effective) of the denotational value of every term of $\mathcal{L}_{\mathsf{TTT}}$.

The formulation of the semantics is based on Gallin's [145] widely-used semantics for many-sorted STT, which was based on *Henkin's semantics* [171], while I mainly extend it to treat the partiality and ramification of TTT.[18] The semantics replaces an informal, and sometimes inexact, description by Tichý [422] (or [416]) of the ways the constructions v-construct. The semantics can also be used for languages different from $\mathcal{L}_{\mathsf{TTT}}$, e.g. for $\mathcal{L}_{*\mathcal{T}}$.

I will utilise a semi-formal metalanguage[19] which employs

Notational convention 11 (Main metalanguage symbols)

$x_{(i)}$	a variable for constructions or non-constructions
c	a variable for constructions
C	a metavariable for constructions
f	a variable for functions (as mappings)
$\_$	represents nothing at all (a semantic 'gap')
$\tau_{(i)}$	a(n arbitrary) type (2.4.4)
v	a(n arbitrary) valuation (2.3.1)

First, I will define the notions of *frame* and *model*, while I will also utilise the notion of the *interpretation function*. They are all necessary to fully comprehend (say) the term "$[\lambda n[3 \div n]]$" and so grasp the intended construction: the term "$\div$" is only unambiguously understood if the interpretation function maps it to (say) the division function operating on $\mathbb{N}^0$, though it could alternatively be mapped to the division function operating on $\mathbb{N}$. In the first case,

---

[18] My semantics was first published in Raclavský [322]. For a comparision, see Thomason's [408] model-theoretic semantics (Henkin models) for Churchian [66] RTT. Muskens (e.g. [267], [270], [269]) elaborated relational (partial) TT based on Henkin models by Orey [280].

[19] Despite its appearance, the metalanguage is fully formalisable within TTT.

"$[\lambda n[3 \div n]]$" represents the construction of a partial (in the latter case: a total) function of dividing the number three by another number.

Definition 26 (Frame, $\mathfrak{F}$)

A *frame* $\mathfrak{F}$ is an n-tuple

$$\mathfrak{F} = \langle \tau_{1B}, \tau_{2B}, ..., \tau_{lB}, *_1, *_2, ..., *_n, \langle \tau_1 \times ... \times \tau_m \rangle \mapsto \tau_0 \rangle$$

where each type is a type over the same base B; $\tau_{1B}, \tau_{2B}, ...\tau_{lB}$, are all members of B; $*_1, *_2, ..., *_n$ are all types of constructions over B; the expression "$\langle \tau_1 \times ... \times \tau_m \rangle \mapsto \tau_0$" stands for all types of i-ary functions over B (for $1 \leqslant i \leqslant m$), everything in accordance with Definition$_{\mathrm{TTT}}$.

Definition 27 (Interpretation function, $\mathfrak{I}$)

An *interpretation function* $\mathfrak{I}$ is a function such that it maps each term "${}^0X$" to the designated object X that is a member of a 1st-order (rarely a kth-order, for $1 < k$) type over B, and $\mathfrak{I}$ naturally extends to other terms in conformity with the grammar of the typed language in question.

Remarks. Typical terms that require explicit determination of denotation by means of $\mathfrak{I}$ are primitive terms such as "${}^{(0)}\neg$", "${}^{(0)}\wedge$", "${}^{(0)}\forall^{\tau}$" and "${}^0 =^{\tau}$" (in this case even for greater orders than 1). The rest of $\mathfrak{I}$ can then be established in a proof-theoretic manner by means of rules; e.g., the meaning of "${}^{(0)}\vee$" is definable with the help of "${}^{(0)}\neg$" and "${}^{(0)}\wedge$". In principle, the method can also be applied to natural language, though the determination of denotation of its primitive expressions creates an obvious obstacle.

Definition 28 (Model, $\mathfrak{M}$)

A model $\mathfrak{M}$ is a couple

$$\mathfrak{M} = \langle \mathfrak{F}, \mathfrak{I} \rangle$$

where $\mathfrak{F}$ is a frame and $\mathfrak{I}$ is an interpretation function.

Convention 12 ('$[\![C]\!]^{\mathfrak{M},v}$')

"$[\![C]\!]^{\mathfrak{M},v}$" is short for "the denotational value of term "C", given $\mathfrak{M}$ and v".

There is also another, equally convenient, reading of "$[\![C]\!]^{\mathfrak{M},v}$":

"the construction C v-constructs".

(Since C itself is already an object of a certain $\mathfrak{M}$, the indication "$\mathfrak{M}$" would be redundant.) In this book, I will often utilise such an 'objectual way of speech': instead of speaking about denotation of terms, I will directly speak about objects the constructions v-construct.

Definition 29 (Semantics for $\mathcal{L}_{\mathsf{TTT}}$)

1. (*variables*)
$[\![V_j^\tau]\!]^{\mathfrak{M},v}$ = the only x such that $sq^\tau \in v$ and $x = sq^\tau(j)$, for $j \in \mathbb{N}$ [20]

2. (*0-executions*)
$[\![{}^0X]\!]^{\mathfrak{M},v} = X$

3. (*1-executions*)
$$[\![{}^1X]\!]^{\mathfrak{M},v} = \begin{cases} [\![C]\!]^{\mathfrak{M},v}, \text{ if } X \text{ is a certain construction } C \text{ and } \exists x(x = [\![C]\!]^{\mathfrak{M},v}) \\ - \end{cases}$$

4. (*2-executions*)
$$[\![{}^2X]\!]^{\mathfrak{M},v} = \begin{cases} [\![[\![C]\!]^{\mathfrak{M},v}]\!]^{\mathfrak{M},v}, \text{ if } X \text{ is a certain construction } C \\ \text{and } \exists c(c = [\![C]\!]^{\mathfrak{M},v} \wedge \exists x(x = [\![c]\!]^{\mathfrak{M},v})) \\ - \end{cases}$$

5. (*compositions*)
$$[\![\,[CC_1...C_m]\,]\!]^{\mathfrak{M},v} = \begin{cases} f(X_1, ..., X_m) \text{ if } [\![C]\!]^{\mathfrak{M},v} = f \in \langle \tau_1 \times ... \times \tau_m \rangle \mapsto \tau, \\ [\![C_1]\!]^{\mathfrak{M},v} = X_1 \in \tau_1, \text{ ..., and } [\![C_m]\!]^{\mathfrak{M},v} = X_m \in \tau_m \\ \text{and } \exists x(x = f(X_1, ..., X_m)) \\ \\ - \end{cases}$$

6. (*closures*)
$[\![\,[\lambda x_1...x_m.C]\,]\!]^{\mathfrak{M},v}$ = the only $f \in \langle \tau_1 \times ... \times \tau_m \rangle \mapsto \tau$ which maps each possible $\langle [\![x_1]\!]^{\mathfrak{M},v(')}, ..., [\![x_m]\!]^{\mathfrak{M},v(')} \rangle$ to $[\![C]\!]^{\mathfrak{M},v(')}$ (if any), where C is a τ-construction, $[\![x_i]\!]^{\mathfrak{M},v(')} \in \tau_i$ and each v' is like v except regarding what it assigns to (some) variables other than $x_1, ..., x_m$

Remark. For comments on v-constructing and examples of constructions see 2.6.3.
Comments.
i. Using the same metalanguage, one can neatly define the notions of v-proper and v-congruent ($\cong$) constructions (2.3.2). In the second definition, the operator *det* rectifies the fact that some constructions are v-improper.

- C is *v-proper* $=_{df}$ $\exists x([\![C]\!]^{\mathfrak{M},v} = x)$
- C_1 is *v-congruent* ($\cong$) with C_2 $=_{df}$ $det([\![C_1]\!]^{\mathfrak{M},v} = [\![C_2]\!]^{\mathfrak{M},v})$, or both C_1 and C_2 are v-improper

[20] V_j^τ is usually written as a small italicised letter.

Definition 30 (Operator *det*)

$$det(C) = \begin{cases} [\![C]\!]^{\mathfrak{M},v} & \text{only if } [\![C]\!]^{\mathfrak{M},v} = \mathtt{T} \text{ or } [\![C]\!]^{\mathfrak{M},v} = \mathtt{F} \\ \mathtt{F} \end{cases}$$

ii. Other semantic notions are definable, esp. *validity* of constructions (logical truth) and semantic (model-theoretic) *logical consequence* ($\models$).

Definition 31 (Validity of constructions)

C is *valid* $=_{df} \forall v([\![C]\!]^{\mathfrak{M},v} = \mathtt{T})$

Definition 32 (Semantic logical consequence, $\models$)

$\{C_1, ..., C_m\} \models C$
$=_{df} \forall v\ \forall c\ ((c \in \{C_1, ..., C_m\}) \rightarrow (det([\![c]\!]^{\mathfrak{M},v} = \mathtt{T}) \rightarrow det([\![C]\!]^{\mathfrak{M},v} = \mathtt{T})))$

Definitions of validity and consequence applicable to $\mathcal{L}_{TTT}$'s terms should rely on the just given definitions.

2.6.1 Subconstructions

Tichý [422] proposed no definition of the notion of *subconstruction.* My following definition takes into account that a. a construction C can contain a subconstruction of an order lower than C's order (this is peculiar to constructions of the form ${}^0X, {}^1X, {}^2X$), and that b. some subconstructions of a certain construction C are not 'activated' during C's v-constructing. Cases in b. fall under cases in a. To illustrate, the kth-order construction ${}^0C$ v-constructs the $(k-1)$st-order construction C, but the fact that C v-constructs something (if it is v-proper) is irrelevant for C's v-constructing; my definition views the 'inactivated' C as a subconstruction of ${}^0C$.

Definition 33 (kth- and $(k-1)$st-order subconstruction)

Let $C_{(i)}$ and variables $x_1, x_2, ..., x_n$ be kth-order constructions, for $1 \leqslant k$, and X be a $(k-j)$th-order non-construction or a $(k-j)$th-order construction, for $1 \leqslant j < k$.

1. C is a *kth-order subconstruction* of C.
2. If X is a $(k-1)$st-order construction, then X is a $(k-1)$*st-order subconstruction* of
   (a) $C_{(i)}$ if $C_{(i)}$ is of form ${}^0X$, ${}^1X$, or ${}^2X$;
   (b) $[CC_1 \ldots C_m]$ and also $[\lambda x_1 \ldots x_m.C]$ if $C_{(i)}$ is their kth-order subconstruction and is of form ${}^0X$, ${}^1X$, or ${}^2X$.

3. $C, C_1, \ldots, C_m$ are *kth-order subconstructions* of $[CC_1 \ldots C_m]$.

4. C is a *kth-order subconstruction* of $[\lambda x_1 \ldots x_m.C]$.

5. Nothing is a kth-*order* or a $(k-1)$st-*order subconstruction* of C, unless it so follows from 1., 2., 3. or 4.

Remark. For example, the 2nd-order subconstructions of the 2nd-order (type-theoretically inconsistent) construction $[^0[3+2] \div 1]$ are $[^0[3+2] \div 1], ^0[3+2], ^{(0)}\div, ^{(0)}1$, and its 1st-order subconstructions are $[3+2], ^{(0)}3, ^{(0)}+, ^{(0)}2$. In conformity with the following agreement, the last four subconstructions are not direct subconstructions of the whole construction.

Convention 13 (Direct subconstruction)

A construction $C_{(i)}$ is called a *direct subconstruction* of a kth-order construction C iff $C_{(i)}$ is C's kth-order subconstruction.

Convention 14 (Occurrence of variable)

A variable x may occur in a construction C repeatedly, which is why I speak about different *occurrences of the variable* x in C.

Remark. The agreement may, of course, be extended for other kinds of constructions, but the present form is sufficient for my following definition of the notion of free variable. Continuing my discussion in 2.3, the binding sequence "$\lambda x_1 ... x_m$" does not contain any occurrence of a variable $x_{(i)}$, since in $[\lambda x_1 ... x_m.C]$, it only indicates closure of C. The definition of the notion of subconstruction thus entails that the variable $x_{(i)}$ may only occur in the 'body' C of $[\lambda x_1 ... x_m.C]$.

2.6.2 Freedom of variables

The following definition of the notion of *free (or: bound) variable* extends the definition from Raclavský, Kuchyňka and Pezlar [339]. By its point 2.(b), it corrects an error in Tichý's original definition [422], according to which e.g. the variable x has even been treated as not free in constructions such as ^{20}x, though ^{20}x is v-congruent with x, which means that x in ^{20}x is obviously free (Raclavský [328]).

Definition 34 (Free variable)

Let x be an arbitrary variable and $C_{(i)}$ a construction.

1. The variable x is *free* in x.

2. (a) If x is free in at least one of its occurrences in C, then x is *free* in ^1C.

(b) If x is free in at least one of its occurrences in C, or in D while C is then of the form ${}^{(1...1)0}D$ (where "(1...1)" indicates iteration of 1-executions), then x is *free* in ${}^{2}C$.

3. If x is free in at least one of its occurrences in $C, C_1, ...,$ or C_m, then x is *free* in $[CC_1...C_m]$.

4. If x is free in at least one of its occurrences in C and is distinct from the variables $x_1, ..., x_m$, then x is *free* in $[\lambda x_1 \dots x_m.C]$.

5. Variable x is not *free* in a construction, unless it so follows from 1., 2., 3. or 4.

Convention 15 (Bound variable)

A(n occurrence of a) variable x is called *bound* in a construction C iff it is not free in C.

Remarks. A construction C is occasionally called a *closed construction* iff C contains no free variable; otherwise, C is called an *open construction*.[21] The variable x is not free e.g. in all its occurrences in $[\lambda x[f\,x]], {}^{0}x, {}^{00}x, {}^{0}[f\,x], {}^{0}[\lambda x[f\,x]]$. The so-called λ-binding of x in (say) $[\lambda x[f\,x]]$ is well known in λ-calculus; λ-bound variables are sometimes called "local variables", the free ones are called "global variables". Binding of x within the scope of 0-execution is entailed by the specification of 0-executions. The variable x is free in all its occurrences in $x, {}^{20}x, [f\,x], {}^{20}[f\,x], [\lambda y.x]$; in $[[\lambda yx[f\,yx]]\,{}^{0}x\,x]$, x is free because of its occurrence that is recorded quite in the right (x is bound in other occurrences).

2.6.3 Comments on v-constructing and examples

Provided it would be stylistically acceptable, I will simplify the wording of my examples utilising the following convention.

Convention 16 ('$\underset{v}{\leadsto}$')

"$C \underset{v}{\leadsto} X$", where C is a construction and X is an object or nothing (which will be denoted by "_"), is short for "C v-constructs X".

i. To illustrate the v-constructing of variables, assume that v_1 is the aforementioned schematic valuation (2.3.1); then, $x_1^{\tau_2} \underset{v_1}{\leadsto} X_1^{\tau_2}$; $x_2^{\tau_2} \underset{v_1}{\leadsto} X_2^{\tau_2}$, etc. Occasionally, I will deploy a familiar way of speaking about variables.

[21] Instead, Tichý [422] used the distinction "complete/incomplete construction". It underlines the fact that a closed construction involves everything one needs to construct an object (if any), but an open construction requires a contribution from valuation.

Convention 17 (Assignment of objects to variables, variable value, variable range)

Instead of "x^τ v-constructs an object X", I will occasionally say "*v assigns* an object X to the variable x^τ". An object X assigned to x^τ by v will occasionally be called the "*value of the variable x^τ on v*". A type τ, whose members are v-constructed by x^τ, will occasionally be called the "*(variable) range* of x^τ".

ii. Executions ${}^0X$, ${}^1X$ and ${}^2X$ each correspond, to some extent, to the ordinary notion of constant. The differences may be adumbrated by the following, somewhat Fregean, consideration: Every construction is a mode of presentation of a certain object (or nothing if the construction is v-improper), while a. ${}^0X$ presents X as it is (without any change of X); b. ${}^1X$ presents an object (if any) that is obtained by an execution of X; c. ${}^2X$ presents an object (if any) which is obtained by an execution of the object (if any) that is obtained by an execution of X; X (as such) differs from the constructions ${}^0X$, ${}^1X$ and ${}^2X$ in that it occurs outside of any presentations, so to speak. For a particular example, consider the variable c that v-constructs constructions of a certain order k; then

${}^0c$	v-constructs (on any v) the construction c
${}^1c$	v-constructs a construction, say C, which is v-constructed by c
${}^2c$	v-constructs (say) a number v-constructed by (say) C which is v-constructed by c

iii. For every object or construction X and an arbitrary $0 \leqslant n$ there exists an n-multiple execution of X. For $2 < n$, every n-execution X can be identified with (i.e. it is reducible to) $(n-1)$-multiple 2-execution of X.[22] E.g. ${}^4X$ is reducible to ${}^{222}X$. On the other hand, 1- and 2-executions are not reducible to other kinds of constructions.

iv. When dealing with his STT [415] Tichý claimed that every objects (say a number) is a trivial construction of itself, so we may imagine that he identified ${}^0X$ with its v-constructed object X. In the case of 1st-order objects (of which STT is consisting), a discrimination between X and ${}^0X$ would indeed seem too fastidious. The situation changes, however, if X of ${}^0X$ is a construction, i.e. with an admission of higher levels of RTT.[23] To illustrate, the variable x v-constructs a certain object, but ${}^0x$ v-constructs x as such, not the object

[22] Kuchyňka first proposed such a definition in Raclavský, Kuchyňka and Pezlar [339].

[23] In our personal communication Oddie pointed out that if constructions are members of our type hierarchy, it is indefensible to say that a construction v-constructs itself – X can never be a construction of itself. In consequence of this, the first real introduction of ${}^0X$ is Tichý [422].

v-constructed by x. ${}^0C$ thus involves 'manipulating' C as such, regardless of what it v-constructs.

For another example, the sentence

"$3 \div 0$ is an abortive procedure."

ascribes a certain property only to $[3 \div 0]$ (the subject of predication is not the gap, which is the result of $[3 \div 0]$'s v-constructing). The logical analysis of the sentence thus utilises 0-execution of this construction, i.e. ${}^0[3 \div 0]$. One need not, therefore, jump to the linguistic level and try to convince readers that the sentence says that it is a particular expression "$3 \div 0$" of the familiar arithmetical language which is the abortive procedure in question.[24]

v. 1-execution of X, i.e. ${}^1X$, is almost indiscernible from X, if X is a construction, because they both v-construct the same object (or they are both v-improper). But unlike ${}^1X \cong X$, the identity ${}^1X = X$ holds only if X is a v-proper construction. Moreover, ${}^1X$ has a greater order than X.[25] Examples: ${}^1x$ v-constructs what is v-constructed by x; ${}^{10}3$ v-constructs what is v-constructed by ^{03}; ${}^13 \underset{v}{\rightsquigarrow}$ _ (is v-improper), for the non-construction 3 cannot be executed.

vi. As with 1-executions, 2-executions form an ineliminable kind of constructions, for they both correspond (in some way) to the word "construct". Examples: ${}^23 \underset{v}{\rightsquigarrow}$ _, for 3 cannot be executed; ${}^{20}C$ v-constructs what is v-constructed by C. In this book, I will often utilise the aforementioned case with ${}^2c$ (where c ranges over a type of constructions) which v-constructs an object (if any) v-constructed by the value of c. The difference between ${}^1X$ and ${}^2X$ is apparent e.g. from the fact that ${}^2X$ is not reducible to ${}^{11}X$, because if X is a construction, ${}^{11}X$ is v-congruent with ${}^1X$, which in turn is v-congruent with X; however, ${}^2X$ is not.

vii. As regards compositions, note that, unlike in λ-calculus, the record "$[CC_1 \ldots C_m]$" neither represents the result of application of a function to an argument,[26] nor does it represent an application of a function to an argument: it is a composition of certain constructions. One can only speak about a representation of a function to an argument if the first construction v-constructs a function applicable to the string of objects v-constructed by the string of constructions $C_1 ... C_m$. An example of composition: $[[2+1] \div n]$ $v(0/n)$-constructs nothing at all, but, on other valuations v', it v'-constructs the result of dividing 3 (v'-constructed by $[2+1]$) by the number v'-constructed by n.

---

[24] The example and argumentation is adopted from Tichý [422].

[25] Tichý [422] wrongly claimed that ${}^1X$ is nothing but X.

[26] Some writers (e.g. Nederpelt and Geuvers [273]) also think that e.g. the term "$[[\lambda x[x \times x]]\ 7]$" is not the result of application of $[\lambda x[x \times x]]$ to 7, since they treat the term as leading to "$[7 \times 7]$". Some writers explicitly denote such an idea by "$[App[f\ x]]$" instead of "$[f\ x]$" (which still presents the result of application of f to x).

Compositions can also be v-improper because their subconstructions need not v-construct i. an m-ary function or ii. a type-theoretically appropriate m-ary string of objects. For example, $[3\,0\div]$ (wrong order of subconstructions), $[3\div]$ (the second part of a possible argument is missing), $[\div \div 3]$ (a part of an argument is type-theoretically inappropriate). Even $[[\lambda x[xx]][\lambda x[xx]]]$ is v-improper because the function v-constructed by $[\lambda x[xx]]$ (whose record is the left-hand part of the record of the whole construction) is not applicable to the object v-constructed by $[\lambda x[xx]]$ (whose record is the right-hand part of the record of the whole construction).

viii. Closure $[\lambda x_1...x_m.C]$ epitomises the idea that one abstracts from a particular valuation which affects at least one of the variables $x_1, ..., x_m$ that typically occur freely in C (i.e. in the 'body' of the closure). This way, one achieves a closure with regards to the variable range of the variable(s) and v-constructs a function from the range. A closure of C can be performed at once (i.e. '$\lambda x_1 x_2...x_m$'), or gradually (i.e. '$\lambda x_1 \lambda x_1...\lambda x_m$'), or in a varied order (e.g. '$\lambda x_2 x_1 x_3...x_m$' or '$\lambda x_2 \lambda x_1 \lambda x_3...\lambda x_m$').

Examples: $[\lambda x.y]$ v-constructs the constant function from the range of x, while its constant value is v-constructed by y; $[\lambda x.x]$ v-constructs the identity function for the range of x; $[\lambda n[3 \div n]]$ v-constructs the partial function of dividing 3 by a number; $[\lambda n_1 n_2[n_1 > n_2]]$ v-constructs the binary numeric relation $>$; $[\lambda n_1[\lambda n_2[n_1 > n_2]]]$ v-constructs the unary numeric function from numbers in the range of n_1 to functions from numbers in the range of n_2 to truth values, in accordance with the fact of whether the numbers are smaller than the first numbers.

Chapter 3

Natural deduction for partial type logic

3.1 Overview of the chapter

TTs are usually equipped with a deduction system, esp. *natural deduction in sequent style* (*ND*; the term "natural deduction in Gentzen sequent style" is also used in literature). Current ND develops the contributions by Gentzen [153, 152], Jaśkowski [190], Prawitz [305]; the research is partly conducted within computer science.[1] See e.g. Tennant [403], Andrews [10], Girard, Taylor and Lafont [156], Troelstra and Schwichtenberg [428], Sørensen and Urzyczyn [384], Negri, von Plato and Ranta [274], Indrzejczak [183], Nederpelt and Geuvers [273], Francez [135], de Quieroz, de Oliviera and Gabbay [313].

It is well-known that the ND rules introducing logical operators 'define' them, while the ND elimination rules perform derivations from such definitions (see e.g. the example at the end of chapter 1). Though the original ND lacked structural rules which were fruitfully deployed in *sequent calculus* (*SC*),[2] contemporary systems of ND often have structural rules. To achieve this, their rules do not operate on mere formulae, but on *sequents* of form

$$\Gamma \Rightarrow \varphi,$$

where the (so-called) *antecedent* Γ is a set of formulae; "$\Rightarrow$" (or "$\vdash$", "$\longrightarrow$") is a metalinguistic symbol of a syntactically conceived consequence; the formula φ is called *succedent*. Here is an example of such a rule:

$$\wedge\text{-I}\ \frac{\Gamma \Rightarrow \varphi \qquad \Gamma \Rightarrow \psi}{\Gamma \Rightarrow \varphi \wedge \psi}$$

---

1 For an introduction to ND see e.g. Indrzejczak [182]; for an overview of deduction systems, see e.g. Sundholm [396].

2 See Gentzen [153] and e.g. Troelstra and Schwichtenberg [428], Negri, von Plato and Ranta [274].

where "______" has the same meaning as "/", which represents the relation of derivability of the final sequent from the sequents on the left of "/". Within current TTs, typical sequents are of form (where "τ" is a type)

$$\Gamma \Rightarrow \varphi : \tau,$$

which enables *type inferences* ("type assignments"). For example, from sequents containing (say) "$\tau_1 \rightarrow \tau_2$" and "$\tau_1$" one may derive a sequent containing "τ_2", which materialises the powerful Curry-Howard correspondence (see 2.2 for references).

Despite embracing sequents, ND is not identical to SC, because no succedent of any ND rule may contain more than (or less than) one formula, while succedents of SC rules are lists (or multisets) of formulae. For example, unlike ND, SC has the Cut Rule at its disposal: $\Gamma_1 \Rightarrow \Delta_1, \varphi; \varphi, \Gamma_2 \Rightarrow \Delta_2 \,/\, \Gamma_1, \Gamma_2 \Rightarrow \Delta_1, \Delta_2$, where Γ_i, Δ_i are sets of formulae. In consequence of this, unlike in ND, SC is endowed with left and right rules, e.g. ($\wedge$-E-left): $\Gamma, \varphi \wedge \psi \Rightarrow \Delta / \Gamma, \varphi, \psi \Rightarrow \Delta$ and ($\wedge$-E-right): $\Gamma \Rightarrow \Delta, \varphi \wedge \psi / \Gamma \Rightarrow \Delta, \varphi; \Gamma \Rightarrow \Delta, \psi$.

For his STT, Tichý developed an ND in sequent style in his unpublished book manuscript [418]. He published its core in Tichý [415]; its application to a particular area of arguments occurs in Oddie and Tichý [278]; an extension focussing on identity and substitution occurs in Tichý [416].

However, Tichý [422] offered no system of deduction for his TTT. An appropriate extension of the system of [415], which covers rules for ${}^0X, {}^1X$ and ${}^2X$ that are peculiar to TTT, was elaborated upon by Kuchyňka in Raclavský, Kuchyňka and Pezlar [339].[3] I call the system

"$\mathrm{ND}_{\mathsf{TTT}}$".

The present chapter is a continuation of the previous chapter 2, it provides proof-theoretic specification of constructions. In section 3.2, I introduce the basic notions of $\mathrm{ND}_{\mathsf{TTT}}$; the so-called matches, sequents and rules will be explained as certain constructions. In section 3.3, I will introduce the main rules of $\mathrm{ND}_{\mathsf{TTT}}$. Substitution will be treated in section 3.4, where I first define the substitution function (3.4.2) and show its use in various constructions and rules, esp. the Rule of Substitutivity of Identicals (3.4.4).

---

3 On the other hand, Duží, Jespersen and Materna [110] neglected Tichý's work on deduction. Duží and Jespersen [107] deploy ND, but without Tichý's deduction system. Tichý's deduction system was studied e.g. in Raclavský [327] (translated version [329]) and then mainly in Raclavský, Kuchyňka and Pezlar [339]. Duží [102] published sequent calculus for their TIL 2010 (i.e. TIL within a version of TTT), but regardless of its name, the system was a slightly extended Tichý's ND for STT: there are no rules for ${}^0X, {}^1X, {}^2X$ and the operator ${}^0Sub$, used in some rules, is undefined.

3.2 Matches, sequents and rules of $\mathbf{ND_{TTT}}$

This section adopts results from Raclavský [329, 324]. I will repeatedly deviate from Tichý's terminology, notation, or even his views, yet I will not usually indicate the differences in the text.

I will introduce the key notions of $\mathrm{ND_{TTT}}$ as certain constructions. This way, I avoid the usual presentation of rules as metalinguistic terms w.r.t. an object language (even Tichý used metalanguage within which he manipulated 1st-order constructions of his STT). Indeed, since the ambition of TTT is to provide a universal, 'all encompassing' framework, there is no reason for keeping rules as linguistic objects that are interpretable in a superordinate framework.

In my proposal, rules are higher-order constructions operating on lower-order constructions. That rules are explained as certain constructions enables an easy verification of their correctness by checking whether the rule in question is a valid construction, i.e. a construction which v-constructs T on every v (2.6).

The key notions of $\mathrm{ND_{TTT}}$ are i. congruence statements called "matches"; they form ii. sequents, which in turn form iii. derivation rules. The defined notions will be illustrated using a simple argument whose validity will be established in $\mathrm{ND_{TTT}}$.

$\mathrm{ND_{TTT}}$ implements Frege's (e.g. [141]) idea of *2D-inference*, because arguments are modelled as sequents, i.e. as 'logical truths', while rules operate on the sequents, not on certain constructions qua formulae, i.e. on mere 'hypotheses'.[4]

Definition 35 (Match, $\mathcal{M}$)

A *match* $\mathcal{M}$ is a construction of form

a. $[{}^0C \cong {}^{00}X]$, or
b. $[{}^0C \cong {}^0x]$,

where the constructions $C, {}^0X$ and x, v-construct an object of the same type, i.e. $C, x, {}^0X/\tau$; ${}^0C, {}^{00}X, {}^0x/*_k$; $\cong$ is a relation of v-congruence between kth-order constructions (see 2.3.2, 2.6), i.e. ${}^{(0)}\cong/(o *_k *_k)$. *Empty matches* are also admissible, they are of form

c. $[\neg[\exists^k \lambda c^k [{}^0C \cong c^k]]]$,

where ${}^{(0)}\neg/(oo); {}^{(0)}\exists^k/(o(o*_k)); c^k/*_k$ (a variable for kth-order constructions).

[4] For discussion of 2D-inference see e.g. Tichý [422], Tichý and Tichý [426], de Quieroz, de Oliviera and Gabbay [313], Pezlar [298].

Matches of form a.–c. will usually be written thus:

a. "$C \cong {}^0X$"
b. "$C \cong x$"
c. "$C \cong \_$"

Remark. Tichý used the notation "$C : x$" and "$C :\ {}^0X$", which might suggest a similarity of matches to typing declarations such as "$t : \tau$". Indeed, if one knows the type of object v-constructed by x (or ${}^0X$), a match codes a certain piece of information about a type related to C; nevertheless, "$t : \tau$" is much more powerful in speaking about types. Thus, a match rather says that C v-constructs and an object v-constructed by x (or ${}^0X$), i.e. it speaks about their v-congruence, which is why I write "$\cong$" instead of ":".

Definition 36 (Satisfaction of match)

A match $\mathcal{M}$ is *satisfied* by a valuation v iff C and ${}^0X$, or C and x, v-construct the same object. An empty match is *satisfied* by v iff the 'schematic match' $C \cong c^k$ contained in the empty match is not satisfied by v.

Notational convention 18 (Construction of a set of constructions)

"$\{C_1, ..., C_n\}$" is short for "$[\lambda c^k[[c^k =^{*_k} {}^0C_1] \vee ... \vee [c^k =^{*_k} {}^0C_n]]]$", where $c^k, {}^0C_1, ..., {}^0C_n / *_k; {}^{(0)}{=}^{*_k} / (o *_k *_k); {}^{(0)}\vee/(ooo)$. In addition, "$\emptyset$" is short for "$[\lambda c^k.\mathrm{F}]$", where ${}^{(0)}\mathrm{F}/o$ (False).

Definition 37 (Sequent, $\mathcal{S}$)

A *sequent* $\mathcal{S}$ is a construction of form

$$[\{\mathcal{M}_1, ..., \mathcal{M}_n\} \Rightarrow \mathcal{M}],$$

where $\mathcal{M}_1, ..., \mathcal{M}_n$ are antecendent matches, $\mathcal{M}_1$ a succedent match, and $\Rightarrow$ is the relation of logical consequence between sets of matches and matches; ${}^0\mathcal{M}_1, ..., {}^0\mathcal{M}_n, {}^0\mathcal{M} / *_{k+1}; {}^0{\Rightarrow} / (o *_{k+1} *_{k+1})$, where k is the order of constructions, whose 0-executions are used in matches $\mathcal{M}_1, ..., \mathcal{M}_n, \mathcal{M}$.

A sequent will usually be written thus:

$$\text{"}\{\mathcal{M}_1, ..., \mathcal{M}_n\} \Rightarrow \mathcal{M}\text{".}$$

Remark. In fact, $\Rightarrow$ is the semantic notion of logical consequence (2.6). The notation change from "$\models$" to "$\Rightarrow$" underlies the fact that, on syntactic considerations within deduction, the semantic character of logical consequence is suppressed.

Notational convention 19 ('Bidirectional' sequent, $\Leftrightarrow$)

"$\mathcal{M}_1 \Leftrightarrow \mathcal{M}_2$" is short for "$\{\mathcal{M}_1\} \Rightarrow \mathcal{M}_2$, $\{\mathcal{M}_2\} \Rightarrow \mathcal{M}_1$".

Definition 38 (Validity of a sequent)

A sequent $\mathcal{S}$ is *valid* iff every valuation v that satisfies all matches forming $\mathcal{S}$'s antecedent also satisfies $\mathcal{S}$'s succedent match.

Remark. A *linguistically formulated argument* (briefly: *linguistic argument*) $\mathcal{A}$ consists of a set of sentences $S_1, ..., S_n$ called *premises* and a sentence S called a *conclusion*, which is separated from $S_1, ..., S_n$ by a word such as "Therefore", or by "________" or "/". Adequate logical analyses (4.2) of $S_1, ..., S_n$ and S, each coupled with '$\cong o$', form a sequent which I consider to be an *argument* in the proper sense of the word (an argument is thus language independent).

Convention 20 (Argument, logical form of an argument)

An *argument* is a sequent whose matches are of form $[{}^0C_{(i)} \cong {}^0o]$, where o is a variable for truth values and ${}^0C_{(i)}$ is a construction of a truth value. The *logical form of an argument* is the sequent that emerged from an argument by a legitimate replacement of 0-executions of extra-logical objects by type-theoretically appropriate variables.

Example. For an illustration consider the linguistically formulated argument

"One plus two makes three. Therefore, one plus two makes three."

The corresponding argument, denoted below by "$U_{1+2=3}$", is the following sequent (an unabbreviated record)

$$[\{{}^0[[[1+2] =^{\nu} 3] \cong o]\} \Rightarrow {}^0[[[1+2] =^{\nu} 3] \cong o]],$$

where ${}^{(0)}1, {}^{(0)}2, {}^{(0)}3/\nu$ (type of numbers $\mathbb{N}^0$); $o/o; {}^{(0)}=^{\nu} /(o\nu\nu)$. The logical form of "$FU_{1+2=3}$" is the following sequent (an unabbreviated record)

$$[\{{}^0[[[n_1 + n_2] =^{\nu} n_3] \cong o]\} \Rightarrow {}^0[[[n_1 + n_2] =^{\nu} n_3] \cong o]],$$

where $n_1, n_2, n_3/\nu; {}^{(0)}+/(\nu\nu\nu)$. In this book, I simplify records of arguments to meet the common form; i.e. I will write only the left parts of matches and "$\Rightarrow$" will be replaced by "________".

Definition 39 (Derivation rule, $\mathcal{R}$)

A *(derivation) rule* $\mathcal{R}$ is a construction of form

$$[\{{}^0\mathcal{S}_1, ..., {}^0\mathcal{S}_m\} \cup \{{}^0\mathcal{P}_1, ..., {}^0\mathcal{P}_n\} \Rightarrow {}^0\mathcal{S}],$$

where $\mathcal{S}_1, ..., \mathcal{S}_m$ are $(k+2)$nd-order sequents and $\mathcal{P}_1, ..., \mathcal{P}_m$ are $(k+3)$rd-order constructions expressing *conditions* which delimit the validity of the rule; i.e. ${}^0\mathcal{S}_1, ..., {}^0\mathcal{S}_m / *_{k+2}; {}^0\mathcal{P}_1, ..., {}^0\mathcal{P}_m / *_{k+3}$, where k is the order of constructions whose 0-executions occur in matches of $\mathcal{S}_1, ..., \mathcal{S}_m, \mathcal{S}$.

A rule $\mathcal{R}$ will usually be written thus (conditions will often be expressed verbally and "$\mathcal{R}$" will often be written aside):

$$\mathcal{R}\,\frac{\mathcal{S}_1 \qquad \mathcal{S}_2 \quad \ldots \quad \mathcal{S}_m}{\mathcal{S}}$$

Condition: $\mathcal{P}_1; \ldots; \mathcal{P}_n$.

Definition 40 (Soundness of a rule)

A rule

$$\mathcal{R}\,\frac{\mathcal{S}_1 \qquad \mathcal{S}_2 \quad \ldots \quad \mathcal{S}_m}{\mathcal{S}}$$

is *sound* iff it holds that if all $\mathcal{R}$'s sequents $\mathcal{S}_1, \ldots, \mathcal{S}_m$ are valid, then $\mathcal{R}$'s sequent $\mathcal{S}$ is also valid.

Convention 21 (Derivability by a rule $\mathcal{R}$, $\vdash_{\mathcal{R}}$)

The final sequent $\mathcal{S}$ of a rule $\mathcal{R}$ will be called (immediately) *derivable from* $\mathcal{R}$'s sequents $\mathcal{S}_1, \ldots, \mathcal{S}_m$ *according to a rule* $\mathcal{R}$, writing it "$\vdash_{\mathcal{R}}$".

Example. To illustrate, consider the Rule of Starting Sequent (AX), i.e. $[[\emptyset \cup \{[^0\mathcal{M} \in M]\}] \Rightarrow {}^0[M \Rightarrow \mathcal{M}]]$ (see its abbreviated record in 3.3.1), according to which its final sequent (which says that the match $\mathcal{M}$ is a logical consequence of the set of matches M containing $\mathcal{M}$) is derivable from the empty set of sequents. Then, $\vdash_{(\mathrm{AX})} U_{1+2=3}$ and $\vdash_{(\mathrm{AX})} FU_{1+2=3}$.

Definition 41 (Derivation $\mathcal{D}$ w.r.t. a set of rules R)

A sequence of sequents is a *derivation* $\mathcal{D}$ of a sequent $\mathcal{S}$ from a set of sequents H *w.r.t. a set of rules* R iff the sequence is finite and each of its sequent is derived from H's members, or from a set of sequents derived thus from H according to a rule $\mathcal{R} \in R$, while $\mathcal{S}$ is the last member of the sequence.

A derivation $\mathcal{D}$ will usually be written thus:

$$\frac{H}{\mathcal{S}}\,R$$

or:

"$\mathcal{S}_1, \ldots, \mathcal{S}_m \vdash_R \mathcal{S}$"

(If R is a singleton, I omit "{", "}", writing simply "$\vdash_{\mathcal{R}}$".)

Example. Consider that, from $H = \{U_{1+2=3}\}$, we derive the sequent $U'_{1+2=3}$ which differs from $U_{1+2=3}$ in containing $[3 =^{\nu} [2+1]]$ instead of $[[1+2] =^{\nu} 3]$. Our R should contain the rules of commutativity of $=^{\nu}$ and $+$. The sequence of two such sequents is a derivation of the final sequent from H w.r.t. R in question: $U_{1+2=3} \vdash_R U'_{1+2=3}$.

Definition 42 (Derived rule, $\mathcal{R}'$)

Let H be a set of sequents, $\mathcal{S}$ be a sequent and $\mathcal{P}_1, ..., \mathcal{P}_m$ be a set of conditions of rules. A construction $\mathcal{R}'$ of form $[H \cup \{{}^0\mathcal{P}_1, ..., {}^0\mathcal{P}_m\} \Rightarrow {}^0\mathcal{S}]$ is a *derived rule w.r.t.* a set of rules R iff $\mathcal{R}' \notin R$ and there exists a derivation of $\mathcal{S}$ from H w.r.t. R (while satisfying conditions $\mathcal{P}_1, ..., \mathcal{P}_m$).

Definition 43 (Argument licensed by a rule)

An *argument* $\mathcal{S}$ is *licensed by a rule* $\mathcal{R}'$ over a set of rules R iff $\mathcal{R}'$ is of form $[\emptyset \cup \{{}^0\mathcal{P}_1, ...{}^0\mathcal{P}_n\} \Rightarrow {}^0\mathcal{S}]$ and $\mathcal{R}' \in R$ or $\mathcal{R}'$ is derived w.r.t. R.

Example. Consider the derivation $\mathcal{D}$ similar to the aforementioned one, whereby one derives the sequent of form $FU_{3=1+2}$ (which contains $[n_3 =^\nu [n_2 + n_1]]$) from $H = \{FU_{1+2=3}\}$ w.r.t. R mentioned above. $\mathcal{D}$ yields a derivation rule $\mathcal{R}'$, whose final sequent is of form $FU_{1+2=3}$. $\mathcal{R}'$ and the rule of substitution (3.4.4) license the argument expressed by "$1 + 2 = 3$. Therefore, $3 = 2 + 1$.".

Convention 22 (Definition)

By a *definition* I will call a derivation rule $\mathcal{M}_1 \Leftrightarrow \mathcal{M}_2$ such that each $\mathcal{M}_i$ is typically of form $C \cong o$ (rarely: $C \cong \_$), where $o, C/o$, while C in $\mathcal{M}_1$ contains the construction introduced and C in $\mathcal{M}_2$ is a compound of certain previously introduced constructions.

Remark. Various writers treat definitions as rules, see esp. works by Schroeder-Heister, e.g. [365], or Nederpelt and Geuvers [273]. Definition as a rule is a notion relative to a *derivation system*, i.e. a system of rules and constructions on which the rules operate.[5] Definitions indicate which notions are derived and which are primitive in a particular derivation system. For example, whether ${}^{(0)}\neg$ and ${}^{(0)}\rightarrow$, or rather ${}^{(0)}\neg$ and ${}^{(0)}\vee$ are primitive in a derivation system.

3.3 Derivation rules of $\mathrm{ND}_{\mathrm{TTT}}$

Derivation rules can be split into three groups: i. structural rules that govern the behaviour of sequents (3.3.1), ii. constructional rules that govern the general behaviour of constructions (3.3.2), iii. logical rules that govern the behaviour of constructions involving logical operators (3.3.3). One may also add iv. rules peculiar to a particular area (e.g. arithmetic).

As I have shown at the end of chapter 1, the rules of groups ii.–iv. proof-theoretically determine the meanings of the key terms involved in them. In particular, the behaviour of constructions – which has been expressed by the

[5] For a precise definition of the notion see Raclavský and Kuchyňka [338] and also 6.2.1.

description of the model-theoretic semantics of the terms representing the constructions (2.6) – can alternatively be expressed in a proof-theoretic manner by the rules of group ii.

In records, I will utilise, among others, the following metalinguistic symbols and the convention stated below. I will assume no mismanagement of orders appears (so if C is a k-order construction, C should v-construct an object belonging to a type of order k).

Notational convention 23 (Arbitrary matches, arbitrary variables, etc.)

$\mathcal{M}_{(i)}, \mathcal{M}'$	arbitrary matches
Γ, Γ'	arbitrary sets of matches
$A, X_{(i)}, Y, F, C$	arbitrary constructions
$a, x_{(i)}, y, f$	arbitrary (pairwise distinct) variables
$\tau_{(i)}$	arbitrary type(s)
$Y_{(X/x)}$	construction Y, in which all free occurrences of x are replaced by ${}^0X$, see the substitution function 3.4.2
$Y_{(\bar{X}_m/\bar{x}_m)}$	Y which results from m (nested) substitutions

Notational convention 24 ('$\tilde{x}$', '$\bar{x}$', '$\mathbf{x}$')

"$\tilde{x}$" (etc.) is short for "$x_1x_2...x_m$".
"$\bar{x}$" (etc.) is short for "$x_1, x_2, ..., x_m$".
"$\mathbf{x}$" (etc.) occurs in an expression that represents both its variant with "x" instead of "$\mathbf{x}$" and with "${}^0X$" instead of "$\mathbf{x}$".

I will also utilise familiar abbreviations such as "$\Gamma, \mathcal{M} \Rightarrow \mathcal{M}'$" in lieu of "$\Gamma \cup \{\mathcal{M}\} \Rightarrow \mathcal{M}'$" and "$\Gamma, \Gamma'$" in lieu of "$\Gamma \cup \Gamma'$". Moreover,[6]

Convention 25 (Patently incompatible matches)

Two matches $\mathcal{M}$ a $\mathcal{M}'$ will be called *patently incompatible matches* iff they are of forms a. $C \cong \mathbf{a}$ and $C \cong \_$, or b. $C \cong \mathbf{a}_1$ and $C \cong \mathbf{a}_2$, where $\mathbf{a}_1, \mathbf{a}_2$ are not v-congruent.

Convention 26 (Variable free in match/set of matches)

A variable x will be called *free in a match* $C \cong \mathbf{a}$ or $C \cong \_$, iff x is free in $\mathbf{a}$ or C. A variable x will be called *free in a set of matches* $\{\mathcal{M}_1, ..., \mathcal{M}_n\}$, iff it is free in at least one of matches $\mathcal{M}_1, ..., \mathcal{M}_n$.

---

[6] Tichý [416] provided details on variable's being free for a type-theoretically specific substitute. I assume extension of his investigation for TTT.

Convention 27 (Variable free in C for D)

A variable x will be called *free in C for D* iff x is free in C and $D, x/\tau$ (i.e. D and x match in their type).

3.3.1 Structural Rules of $\mathbf{ND_{TTT}}$

Definition 44 (Structural rules)

The Rule of Starting Sequent

$$(\text{AX})\ \frac{}{\Gamma, \mathcal{M} \Rightarrow \mathcal{M}}$$

Remark. This rule serves as the well-known 'start axiom' $A \Rightarrow A$. Tichý used a slightly different formulation of the rule and called it "the Rule of Trivial Sequent" (TS).

The Rule of Weakening

$$(\text{WR})\ \frac{\Gamma \Rightarrow \mathcal{M}}{\Gamma', \Gamma \Rightarrow \mathcal{M}}$$

Remark. This rule serves as the Rule of Weakening (Thinning) of SC (Tichý called it the Rule of Redundant Match, RM). Note that we use matches and their sets, not multisets as in SC.

The Rule of Simplification

$$(\text{SIM})\ \frac{\Gamma, \mathcal{M}' \Rightarrow \mathcal{M} \qquad \Gamma \Rightarrow \mathcal{M}'}{\Gamma \Rightarrow \mathcal{M}}$$

Remark. This rule is a 'metalinguistic' Modus ponens. Note also its loose resemblance to the Rule of Cut.

The Rule of Trivial Match

$$(\text{TM})\ \frac{}{\Gamma \Rightarrow \mathbf{a} \cong \mathbf{a}}$$

Remark. (TM) states v-properness of 0-executions and variables.

The Rule of Vacuous Sequent

$$(\text{VAC})\ \frac{\Gamma \Rightarrow \mathcal{M}_1 \qquad \Gamma \Rightarrow \mathcal{M}_2}{\Gamma \Rightarrow \mathcal{M}}$$

Condition: The matches $\mathcal{M}_1$ and $\mathcal{M}_2$ are patently incompatible.

Remark. (VAC) is a 'metalinguistic' form of *ex contradictione quodlibet.*

The Rule of Exhaustion

$$\text{(EXH)}\ \frac{\Gamma, A \cong \_ \Rightarrow \mathcal{M} \qquad \Gamma, A \cong a \Rightarrow \mathcal{M}}{\Gamma \Rightarrow \mathcal{M}}$$

Condition: a is not free in $A, \mathcal{M}, \Gamma$; $a, A/\tau$.[7]

Remark. (EXH) says that $\mathcal{M}$ follows from Γ even without assuming which object (if any) A v-constructs (i.e. that A is v-proper).

3.3.2 Constructional Rules of ND_{TTT}

Definition 45 (Comp-rules)

The Rule of a-*Substitution*

$$\text{(I)}\ \text{(a-SUB)}\ \frac{\Gamma \Rightarrow [F\tilde{X}_m] \cong \mathbf{y} \qquad \Gamma \Rightarrow X_1 \cong \mathbf{x}_1; \ldots; \Gamma \Rightarrow X_m \cong \mathbf{x}_m}{\Gamma \Rightarrow [F\tilde{\mathbf{x}}_m] \cong \mathbf{y}}$$

$$\text{(II)}\ \text{(a-SUB)}\ \frac{\Gamma \Rightarrow [F\tilde{\mathbf{x}}_m] \cong \mathbf{y} \qquad \Gamma \Rightarrow X_1 \cong \mathbf{x}_1; \ldots; \Gamma \Rightarrow X_m \cong \mathbf{x}_m}{\Gamma \Rightarrow [F\tilde{X}_m] \cong \mathbf{y}}$$

Condition: $\bar{x}_m$ and y are not free in $[F\tilde{X}_m], \Gamma$. $\mathbf{y}/\tau; \mathbf{x}_1, X_1/\tau_1; \ldots; \mathbf{x}_m, X_m/\tau_m; F/(\tau\tilde{\tau}_m)$.

The Rule of Extensionality

$$\text{(EXT)}\ \frac{\Gamma, [\mathbf{f}\tilde{x}_m] \cong y \Rightarrow [\mathbf{g}\tilde{x}_m] \cong y \qquad \Gamma, [\mathbf{g}\tilde{x}_m] \cong y \Rightarrow [\mathbf{f}\tilde{x}_m] \cong y}{\Gamma \Rightarrow \mathbf{g} \cong \mathbf{f}}$$

Condition: $\bar{x}_m$ and y are not free in $\mathbf{f}, \mathbf{g}, \Gamma$.
$y/\tau; x_1/\tau_1; \ldots; x_m/\tau_m; \mathbf{f}, \mathbf{g}/(\tau\tilde{\tau}_m)$.

The Rule of a-*Instantiation*

$$\text{(a-INST)}\ \frac{\Gamma \Rightarrow [F\tilde{X}_m] \cong \mathbf{y} \qquad \Gamma, F \cong f, X_1 \cong x_1, \ldots, X_m \cong x_m \Rightarrow \mathcal{M}}{\Gamma \Rightarrow \mathcal{M}}$$

Condition: $y, \bar{x}_m$ and f are not free in $[F\tilde{X}_m], \mathcal{M}, \Gamma$. $\mathbf{y}/\tau; x_1, X_1/\tau_1; \ldots; x_m, X_m/\tau_m;\ f, F/(\tau\tilde{\tau}_m)$.

[7] For a proof of correctness of (EXH), but also (EXT), (β-CON) and (β-EXP) introduced below, see Tichý [415].

Remark. (a-SUB) says that v-congruent constructions $\mathbf{x}_i$ and X_i are intersubstitutable in compositions. (EXT) expresses derivability of a match that consists of constructions of extensionally identical functions. (a-INST) says that $\mathcal{M}$ follows from Γ without assuming that F and $\bar{X}_m$ are v-proper since this follows from v-properness of $[F\tilde{X}_m]$, which already follows from Γ.

Definition 46 (Clos-rules)

The Rule of Contraction

$$(\beta\text{-CON})\ \frac{\Gamma \Rightarrow [[\lambda\tilde{x}_m.Y]\tilde{X}_m] \cong \mathbf{y}}{\Gamma \Rightarrow Y_{(\bar{X}_m/\bar{x}_m)} \cong \mathbf{y}}$$

Condition: y is not free in $\bar{X}_m, Y, \Gamma$. $\mathbf{y}, Y/\tau; x_1, X_1/\tau_1; ...; x_m, X_m/\tau_m$. For $1 \leqslant i \leqslant m$, each variable in X_i is free in Y for x_i.

The Rule of Expansion

$$(\beta\text{-EXP})\ \frac{\Gamma \Rightarrow X_1 \cong \mathbf{x}_1\ ;\ ...\ ;\ \Gamma \Rightarrow X_m \cong \mathbf{x}_m \qquad \Gamma \Rightarrow Y_{(\bar{X}_m/\bar{x}_m)} \cong \mathbf{y}}{\Gamma \Rightarrow [[\lambda\tilde{x}_m.Y]\tilde{X}_m] \cong \mathbf{y}}$$

Condition: y is not free in $\bar{X}_m, Y, \Gamma$. $\mathbf{y}, Y/\tau; x_1, \mathbf{x}_1, X_1/\tau_1; ...; x_m, \mathbf{x}_m, X_m/\tau_m$. For $1 \leqslant i \leqslant m$, each variable in X_i is free in Y for x_i.

The Rule of λ-Instantiation

$$(\lambda\text{-INST})\ \frac{\Gamma, [\lambda\tilde{x}_m.Y] \cong f \Rightarrow \mathcal{M}}{\Gamma \Rightarrow \mathcal{M}}$$

Condition: f is not free in $[\lambda\tilde{x}_m.Y], \mathcal{M}, \Gamma$. $Y/\tau; x_1/\tau_1; ...; x_m/\tau_m; f/(\tau\tilde{\tau}_m)$.

Remark. (λ-INST) says that $\mathcal{M}$ follows from Γ without assuming that a closure is v-proper. (β-CON) and (β-EXP) are Tichý's version of the famous rules known as β-reduction-by-name and β-expansion; partiality is carefully treated. Kuchyňka constructed a derivation leading from Tichý's β-reduction-by-name to β-reduction-by-value, and even η-reduction rule, see Raclavský, Kuchyňka and Pezlar [339].

Definition 47 (The rules of η-conversion)

$$(\eta\text{-CON})\ \frac{\Gamma \Rightarrow F \cong \mathbf{f} \qquad \Gamma \Rightarrow [\lambda\tilde{x}_m[F\tilde{x}_m]] \cong \mathbf{g}}{\Gamma \Rightarrow F \cong \mathbf{g}}$$

$$(\eta\text{-EXP})\ \frac{\Gamma \Rightarrow F \cong \mathbf{f}}{\Gamma \Rightarrow [\lambda\tilde{x}_m\,[F\tilde{x}_m]] \cong \mathbf{f}}$$

Condition (η-CON, η-EXP): $\bar{x}_m, f$ and g are not free in F, Γ. $x_1/\tau_1; \ldots; x_m/\tau_m; F, \mathbf{f}, \mathbf{g}/(\tau\tilde{\tau}_m)$.

Definition 48 (0-exec-rules)

The Rule of 0-Identity

$$(^0\text{-ID})\ \frac{\Gamma \Rightarrow {}^0X \cong \mathbf{x} \qquad \Gamma \Rightarrow {}^0Y \cong \mathbf{x} \qquad \Gamma \Rightarrow X \cong \mathbf{y}}{\Gamma \Rightarrow Y \cong \mathbf{y}}$$

Condition: x is not free in X, Y, Γ. $\mathbf{x}/*_k; \mathbf{y}, Y, X/\tau$ (where τ is a kth-order type).

The Rule of 0-Vacuous Sequent

$$(^0\text{-VAC})\ \frac{\Gamma \Rightarrow X \cong \mathbf{x} \qquad \Gamma \Rightarrow {}^0Y \cong \mathbf{x}}{\Gamma \Rightarrow \mathcal{M}}$$

Condition: X is a subconstruction of Y. x is not free in $X, \mathcal{M}, \Gamma$. $\mathbf{x}, X, {}^0Y/\tau$.

The Rule of 0-Instantiation

$$(^0\text{-INST})\ \frac{\Gamma, {}^0X \cong x \Rightarrow \mathcal{M}}{\Gamma \Rightarrow \mathcal{M}}$$

Condition: x is not free in $\mathcal{M}, \Gamma$. $x, {}^0X/\tau$.

Remark. (^{0}-ID) says that for any object X, there is just one ${}^0X$. (^{0}-VAC) says that anything follows if ${}^0Y$ v-constructs the same object as its proper subconstruction X. (^{0}-INST) says that $\mathcal{M}$ follows from Γ even without assuming that ${}^0X$ is v-proper.

Definition 49 (1-exec-rules)

The Rules of 1-Introduction

$$(^1\text{-I})\ \frac{\Gamma \Rightarrow X \cong \mathbf{x}}{\Gamma \Rightarrow {}^1X \cong \mathbf{x}} \qquad\qquad (^1\text{-I})\ \frac{\Gamma \Rightarrow X \cong \_}{\Gamma \Rightarrow {}^1X \cong \_}$$

The Rule of 1-Elimination

$$(^1\text{-E})\ \frac{\Gamma \Rightarrow {}^1X \cong \mathbf{x}}{\Gamma \Rightarrow X \cong \mathbf{x}}$$

Condition (^{1}-I, ^{1}-E): x is not free in X, Γ. $\mathbf{x}, X/\tau$.

The Rule of 1-Instantiation

$$(^1\text{-INST})\ \frac{\Gamma \Rightarrow {}^1X \cong \mathbf{x} \qquad \Gamma, X \cong y \Rightarrow \mathcal{M}}{\Gamma \Rightarrow \mathcal{M}}$$

Condition (^{1}-INST): x and y are not free in $X, \mathcal{M}, \Gamma$. $\mathbf{x}, y, X/\tau$.

Definition 50 (2-exec-rules)

The Rule of 2-Introduction

$$(^2\text{-I})\ \frac{\Gamma \Rightarrow X \cong \mathbf{x} \qquad \Gamma \Rightarrow\ {}^0Y \cong \mathbf{x} \qquad \Gamma \Rightarrow Y \cong \mathbf{y}}{\Gamma \Rightarrow\ {}^2X \cong \mathbf{y}}$$

The Rule of 2-Elimination

$$(^2\text{-E})\ \frac{\Gamma \Rightarrow X \cong \mathbf{x} \qquad \Gamma \Rightarrow\ {}^0Y \cong \mathbf{x} \qquad \Gamma \Rightarrow\ {}^2X \cong \mathbf{y}}{\Gamma \Rightarrow Y \cong \mathbf{y}}$$

Condition (^{2}-I, ^{2}-E): x and y are not free in X, Y, Γ. $\mathbf{x}, X/*_k; \mathbf{y}, Y/\tau$ (where τ is a kth-order type).

The Rule of 2-Instantiation

$$(^2\text{-INST})\ \frac{\Gamma \Rightarrow\ {}^2X \cong y \qquad \Gamma, X \cong \mathbf{x} \Rightarrow \mathcal{M}}{\Gamma \Rightarrow \mathcal{M}}$$

Condition: x and y are not free in $X, \mathcal{M}, \Gamma$. $\mathbf{x}, X/*_k; y/\tau$ (where τ is a kth-order type).

Remark. (^{2}-I) and (^{2}-E) show how to introduce or eliminate ^2X when X v-constructs a v-proper construction Y. (^{2}-INST) says that, if ^2X is v-proper, $\mathcal{M}$ follows from Γ even without assuming that X is v-proper.

3.3.3 Logical Rules of $\mathbf{ND_{TTT}}$

Definition 51 (Logical rules – negation)

The Rule of ¬-Introduction

$$(\neg\text{-I})\ \frac{\Gamma, \mathbf{j} \cong \mathbf{i} \Rightarrow \mathcal{M}_1 \qquad \Gamma, \mathbf{j} \cong \mathbf{i} \Rightarrow \mathcal{M}_2}{\Gamma \Rightarrow [\neg \mathbf{j}] \cong \mathbf{i}}$$

Condition: The matches $\mathcal{M}_1$ a $\mathcal{M}_2$ are patently incompatible. i is not free in $\mathbf{j}, \mathcal{M}_i, \Gamma$. $\mathbf{i}, \mathbf{j}/o; {}^{(0)}\neg/(oo)$.

The Rule of Redundant Assumption

$$(\text{RA})\ \frac{\Gamma, \mathtt{T} \cong \mathbf{i} \Rightarrow \mathcal{M} \qquad \Gamma, \mathtt{F} \cong \mathbf{i} \Rightarrow \mathcal{M}}{\Gamma \Rightarrow \mathcal{M}}$$

Condition: i is not free in $\mathcal{M}, \Gamma$. $\mathbf{i}, {}^{(0)}\mathtt{T}, {}^{(0)}\mathtt{F}/o$.

The Rule of ¬-Instantiation

$$(\neg\text{-INST})\ \frac{\Gamma, [\neg \mathbf{i}] \cong i \Rightarrow \mathcal{M}}{\Gamma \Rightarrow \mathcal{M}}$$

Condition: i is not free in $\mathcal{M}, \Gamma$. $i, \mathbf{i}/o; {}^{(0)}\neg/(oo)$.

Definition 52 (Logical rules – implication)

The Rule of →-Introduction

$$(\to\text{-I})\ \frac{\Gamma, \mathbf{i} \cong \mathrm{T} \Rightarrow \mathbf{j} \cong \mathrm{T}}{\Gamma \Rightarrow [\mathbf{i} \to \mathbf{j}] \cong \mathrm{T}}$$

Condition: j is not free in $\mathbf{i}$. $\mathbf{i}, \mathbf{j},{}^{(0)}\,\mathrm{T}/o;{}^{(0)}\!\to /(ooo)$.

The Rule of →-Elimination

$$(\to\text{-E})\ \frac{\Gamma \Rightarrow [I \to J] \cong \mathrm{T} \qquad \Gamma \Rightarrow I \cong \mathrm{T}}{\Gamma \Rightarrow J \cong \mathrm{T}}$$

Condition: $I, J,{}^{(0)}\mathrm{T}/o;{}^{(0)}\!\to /(ooo)$.

The Rule of →-Instantiation

$$(\to\text{-INST})\ \frac{\Gamma, [\mathbf{i} \to \mathbf{j}] \cong i \Rightarrow \mathcal{M}}{\Gamma \Rightarrow \mathcal{M}}$$

Condition: i is not free in $\mathbf{j}, \mathcal{M}, \Gamma$. $i, \mathbf{i}, \mathbf{j}/o;{}^{(0)}\!\to /(ooo)$.

Definition 53 (Logical rules – universal quantifier)

The Rule of ∀-Introduction

$$(\forall\text{-I})\ \frac{\Gamma \Rightarrow [F\,x] \cong \mathrm{T}}{\Gamma \Rightarrow [\forall^{\tau} F] \cong \mathrm{T}}$$

Condition: x is not free in F, Γ. $x/\tau; F/(o\tau);{}^{(0)}\mathrm{T}/o;{}^{(0)}\forall^{\tau}/(o(o\tau))$.

The Rule of ∀-Elimination

$$(\forall\text{-E})\ \frac{\Gamma \Rightarrow [\forall^{\tau} F] \cong \mathrm{T}}{\Gamma \Rightarrow [F\,\mathbf{x}] \cong \mathrm{T}}$$

Condition: x is not free in F, Γ. $\mathbf{x}/\tau; F/(o\tau);{}^{(0)}\mathrm{T}/o;{}^{(0)}\forall^{\tau}/(o(o\tau))$.

The Rule of ∀-Instantiation

$$(\forall\text{-INST})\ \frac{\Gamma, [\forall^{\tau}\,\mathbf{f}] \cong i \Rightarrow \mathcal{M}}{\Gamma \Rightarrow \mathcal{M}}$$

Condition: i is not free in $\mathbf{f}, \mathcal{M}, \Gamma$. $i/o; \mathbf{f}/(o\tau);{}^{(0)}\forall^{\tau}/(o(o\tau))$.

Remark. A common variant of (∀-E) is known as the rule of *Universal Instantiation.*

Definition 54 (Logical rules – existential quantifier)

The Rule of ∃-Introduction

$$(\exists\text{-I})\ \frac{\Gamma \Rightarrow [F\,X] \cong \mathtt{T}}{\Gamma \Rightarrow [\exists^\tau F] \cong \mathtt{T}}$$

Condition: $X/\tau; F/(o\tau);^{(0)}\mathtt{T}/o;^{(0)}\exists^\tau/(o(o\tau))$.

The Rule of ∃-Elimination

$$(\exists\text{-E})\ \frac{\Gamma \Rightarrow [\exists^\tau F] \cong \mathtt{T} \qquad \Gamma, [F\,x] \cong \mathtt{T} \Rightarrow \mathcal{M}}{\Gamma \Rightarrow \mathcal{M}}$$

Condition: x is not free in $F, \mathcal{M}, \Gamma$. $x/\tau; F/(o\tau);^{(0)}\mathtt{T}/o;$ $^{(0)}\exists^\tau/(o(o\tau))$.

The Rule of ∃-Instantiation

$$(\exists\text{-INST})\ \frac{\Gamma, [\exists^\tau \mathbf{f}] \cong i \Rightarrow \mathcal{M}}{\Gamma \Rightarrow \mathcal{M}}$$

Condition: i is not free in $\mathbf{f}, \mathcal{M}, \Gamma$. $i/o; \mathbf{f}/(o\tau);^{(0)}\exists^\tau/(o(o\tau))$.

Remark. Each of the 'quantifiers' $\forall^\tau$ and $\exists^\tau$ is equipped with its own set of rules because the classical Rule for Exchange of Quantifiers – and also many other classical laws – do not behave properly if partiality is admitted.[8] In the case of the Rule for Exchange of Quantifiers, for example, $\exists^\tau$ maps any partial non-empty set (v-constructed by the construction F) to the truth value $\mathtt{T}$, yet $\forall^\tau$ maps the set, as well as the set complementary to it, to $\mathtt{F}$; v-congruence of $[\neg[\forall^\tau F]]$ with $[\exists^\tau[\lambda x[\neg[F\,x]]]$ is thus lost, and so $\forall^\tau$ and $\exists^\tau$ are not interdefinable. To avoid such an undesirable behaviour, logical laws must be rectified by inserting the 'totalising operator' Tot, see its definition right below. For example, the corrected Rule for Exchange of Quantifiers is $[\neg[\forall^\tau[\mathrm{Tot}^{(o\tau)} F]]] \cong o \Leftrightarrow [\exists^\tau[\lambda x[\neg[F\,x]]]] \cong o$.

Definition 55 (Operator Tot)

$\models [[\mathrm{Tot}^{(o\tau)}\ f]x] \Leftrightarrow_o [\exists^\tau[\lambda o'[[[f\,x] =^o o'] \wedge [o' =^o \mathtt{T}]]]]$

where $o, o',^{(0)}\mathtt{T}/o; x/\tau; f/(o\tau);^{(0)}{=^o}/(ooo);^{(0)}\mathrm{Tot}^{(o\tau)}/((o\tau)(o\tau));^{(0)}\wedge/(ooo)$.

Remarks. With regards to logical connectives such as $\wedge$ and $\rightarrow$, $\mathrm{ND_{TTT}}$ behaves as in *Bochvar* [45] *3V-* (i.e. three-valued) *logic* (also called *Kleene weak 3V-logic*): constructions whose main logical operator is a construction of a truth function are v-improper, if the function does not obtain a suitable argument. On the other hand, 'quantifiers' $\forall^\tau$ and $\exists^\tau$ behave as in *Kleene strong 3V-logic* (Kleene [212]): they return a truth value even if they are applied to a partial set.

[8] See e.g. Duží, Jespersen and Materna [110], Raclavský [330, 323], Raclavský, Kuchyňka and Pezlar [339].

Definition 56 (Logical rules – identity)

The Rule of =-Introduction

$$(=\text{-I})\ \frac{\Gamma \Rightarrow X \cong \mathbf{x}}{\Gamma \Rightarrow [\mathbf{x} =^{\tau} X] \cong \mathtt{T}}$$

The Rule of =-Elimination

$$(=\text{-E})\ \frac{\Gamma \Rightarrow [\mathbf{x} \ =^{\tau} \ X] \cong \mathtt{T}}{\Gamma \Rightarrow X \cong \mathbf{x}}$$

Condition (=-I, =-E): x is not free in X, Γ. $\mathbf{x}, X/\tau;^{(0)}\mathtt{T}/o;^{(0)}=^{\tau}/(o\tau\tau)$.

The Rule of =-Instantiation

$$(=\text{-INST})\ \frac{\Gamma, [\mathbf{x} =^{\tau} \ \mathbf{y}] \cong i \Rightarrow \mathcal{M}}{\Gamma \Rightarrow \mathcal{M}}$$

Condition: i is not free in $\mathbf{x}, \mathbf{y}, \mathcal{M}, \Gamma$. $i/o; \mathbf{x}, \mathbf{y}/\tau;^{(0)}=^{\tau} /(o\tau\tau)$.

Definition 57 (Logical rules – trivialisation function)

The Rule of $\textsc{Triv}^{(*_k\tau)}$*-Introduction*

$$(\textsc{Triv}^{(*_k\tau)}\text{-I})\ \frac{\Gamma \Rightarrow A \cong \mathbf{x} \qquad \Gamma \Rightarrow {}^{0}\mathbf{x} \cong \mathbf{a}}{\Gamma \Rightarrow [{}^{0}\textsc{Triv}^{(*_k\tau)}\ A] \cong \mathbf{a}}$$

The Rule of $\textsc{Triv}^{(*_k\tau)}$*-Elimination*

$$(\textsc{Triv}^{(*_k\tau)}\text{-E})\ \frac{\Gamma \Rightarrow [{}^{0}\textsc{Triv}^{(*_k\tau)}\ A] \cong \mathbf{a} \qquad \Gamma \Rightarrow {}^{0}\mathbf{x} \cong \mathbf{a}}{\Gamma \Rightarrow A \cong \mathbf{x}}$$

The Rule of $\textsc{Triv}^{(*_k\tau)}$*-Instantiation*

$$(\textsc{Triv}^{(*_k\tau)}\text{-INST})\ \frac{\Gamma, A \cong \mathbf{x} \Rightarrow \mathcal{M} \qquad \Gamma \Rightarrow [{}^{0}\textsc{Triv}^{(*_k\tau)}\ A] \cong \mathbf{a}}{\Gamma \Rightarrow \mathcal{M}}$$

Condition ($\textsc{Triv}^{(*_k\tau)}$-I, $\textsc{Triv}^{(*_k\tau)}$-E, $\textsc{Triv}^{(*_k\tau)}$-INST): x and a are not free in A, Γ and $\mathcal{M}$ (only for $\textsc{Triv}^{(*_k\tau)}$-INST). $\mathbf{a}/*_k; \mathbf{x}, A/\tau$ (where τ is a kth-order type);${}^{(0)}\textsc{Triv}^{(*_k\tau)}/(*_k\tau)$.

Remark. The rules for $\textsc{Triv}^{(*_k\tau)}$ (see 2.5.1), proposed by Kuchyňka in Raclavský, Kuchyňka and Pezlar [339], are not strictly logical ones, for they are needed in proof of β-reduction-by-value from β-reduction-by-name, i.e. for the proper type-theoretic part of TTT.

Definition 58 (Logical rules – singularisation function)

The Rule of $\imath^\tau$-Introduction

$$(\imath^\tau\text{-I})\ \frac{\Gamma \Rightarrow [F\,\mathbf{x}] \cong \mathtt{T} \qquad \Gamma, [F\,\mathbf{x}] \cong \mathtt{T} \Rightarrow y \cong \mathbf{x}}{\Gamma \Rightarrow [\imath^\tau\, F] \cong \mathbf{x}}$$

Condition: x and y are not free in F, Γ. $\mathbf{x}, y/\tau;^{(0)}\,\mathtt{T}/o; F/(o\tau);^{(0)}\imath^\tau/(\tau(o\tau))$.

The Rule of $\imath^\tau$-Elimination

$$(\imath^\tau\text{-E})\ \frac{\Gamma \Rightarrow [\imath^\tau\, F] \cong \mathbf{x}}{\Gamma \Rightarrow [F\,\mathbf{x}] \cong \mathtt{T}}$$

Condition: x is not free in F, Γ. $\mathbf{x}/\tau;^{(0)}\,\mathtt{T}/o; F/(o\tau);^{(0)}\imath^\tau/(\tau(o\tau))$.

The Rule of $\imath^\tau$-Instantiation

$$(\imath^\tau\text{-INST})\ \frac{\Gamma, [F\,\mathbf{x}] \cong \mathtt{T} \Rightarrow \mathcal{M} \qquad \Gamma \Rightarrow [\imath^\tau\, F] \cong \mathbf{x}}{\Gamma \Rightarrow \mathcal{M}}$$

Condition: x is not free in $F, \mathcal{M}, \Gamma$. $\mathbf{x}/\tau;^{(0)}\,\mathtt{T}/o; F/(o\tau);^{(0)}\imath^\tau/(\tau(o\tau))$.

Remark. The rules ($\imath^\tau$-I) and ($\imath^\tau$-E) are adjusted from Tichý [418]. Note that the usual uniqueness condition for the descriptive operator $\imath x$, i.e. $\forall y(F(y) \rightarrow (y = x))$, is inbuilt in ($\imath^\tau$-I) without using $\forall, \rightarrow, =$ (and an adjustment of $F(y)$ because of partiality).

3.3.4 Controlling validity of inferences

This section offers three illustrations of establishing an argument's validity using a syntactic method, namely $\mathrm{ND}_{\mathsf{TTT}}$. (Cf. also our examples in section 3.2.)

The simplest way how to check validity of a linguistic argument is to formalise it to obtain the argument A, and to check whether A is an instance of some valid derivation rule. More precisely, we try to derive a sequent S whose antecedent only consists of the matches that couple (the logical forms of) A's premisses with '$\cong \mathtt{T}$' and whose succedent only consists of the match that couples (the logical form of) A's conclusion with '$\cong \mathtt{T}$'. ("Γ," is suppressed in all examples.)

Example 15 (Establishing an argument's validity I)

Argument, where $P, Q/o$ (P and Q are arbitrary o-constructions of a given order k); $^{(0)}\rightarrow /(ooo)$ (material conditional):

$$\begin{array}{c} P \\ [P \to Q] \\ \hline Q \end{array}$$

Derivation of the sequent $\mathcal{M}_1, \mathcal{M}_2 \Rightarrow Q \cong \mathrm{T}$, where $\mathcal{M}_1 := P \cong \mathrm{T}$ and $\mathcal{M}_2 := [P \to Q] \cong \mathrm{T}$:

$$\dfrac{\dfrac{\dfrac{}{\mathcal{M}_1 \Rightarrow P \cong \mathrm{T}}\,(\mathrm{AX})}{\mathcal{M}_1, \mathcal{M}_2 \Rightarrow P \cong \mathrm{T}}\,(\mathrm{WR}) \qquad \dfrac{\dfrac{}{\mathcal{M}_2 \Rightarrow [P \to Q] \cong \mathrm{T}}\,(\mathrm{AX})}{\mathcal{M}_1, \mathcal{M}_2 \Rightarrow [P \to Q] \cong \mathrm{T}}\,(\mathrm{WR})}{\mathcal{M}_1, \mathcal{M}_2 \Rightarrow Q \cong \mathrm{T}}\,(\to\text{-E})$$

Remark. Note that every step in each branch of the tree presentation of the derivation is a logical truth, since our approach to inference is two-dimensional (cf. 3.2). Similarly below.

Our second example shows that an argument's validity can be established even if we do not yet have a rule of which the argument is an instance. In that case, we derive the argument's conclusion from the set of its premisses directly, using the rules we already have. We thus establish an instance of a new derived rule that licenses the argument in question.

Example 16 (Establishing an argument's validity II)

Argument, where P, Q and $\to$ are as in the above example:

$$\begin{array}{c} P \\ [P \to Q] \\ \hline [P \to [[P \to Q] \to Q]] \end{array}$$

Derivation of the sequent $\mathcal{M}_1, \mathcal{M}_2 \Rightarrow [P \to [[P \to Q] \to Q]] \cong \mathrm{T}$, where $\mathcal{M}_1 := P \cong \mathrm{T}$, $\mathcal{M}_2 := [P \to Q] \cong \mathrm{T}$:

$$\dfrac{\dfrac{\dfrac{\dfrac{\dfrac{\dfrac{}{\mathcal{M}_1 \Rightarrow P \cong \mathrm{T}}\,(\mathrm{AX})}{\mathcal{M}_1, \mathcal{M}_2 \Rightarrow P \cong \mathrm{T}}\,(\mathrm{WR}) \qquad \dfrac{\dfrac{}{\mathcal{M}_2 \Rightarrow [P \to Q] \cong \mathrm{T}}\,(\mathrm{AX})}{\mathcal{M}_1, \mathcal{M}_2 \Rightarrow [P \to Q] \cong \mathrm{T}}\,(\mathrm{WR})}{\mathcal{M}_1, \mathcal{M}_2 \Rightarrow Q \cong \mathrm{T}}\,(\to\text{-E})}{\mathcal{M}_1 \Rightarrow [[P \to Q] \to Q] \cong \mathrm{T}}\,(\to\text{-I})}{\emptyset \Rightarrow [P \to [[P \to Q] \to Q]] \cong \mathrm{T}}\,(\to\text{-I})}{\mathcal{M}_1, \mathcal{M}_2 \Rightarrow [P \to [[P \to Q] \to Q]] \cong \mathrm{T}}\,(\mathrm{WR})$$

Our last example is similar in spirit, but the argument's conclusion is not a (logical) validity as in the above example.

Example 17 (Establishing an argument's validity III)

Linguistic argument:

$$\frac{\begin{array}{c}\text{“Every prime is divisible by exactly 2 numbers (D2N).”}\\ \text{“7 is a prime.”}\end{array}}{\text{“7 is divisible by exactly 2 numbers.”}}$$

Argument (where $n, {}^{(0)}7/\nu$ (ν is type of natural numbers); ${}^{(0)}\mathrm{Pr}, {}^{(0)}\mathrm{D2N}/(o\nu)$):

$$\frac{\begin{array}{c}[\forall^{\nu}\lambda n[[\mathrm{Pr}\, n] \rightarrow [\mathrm{D2N}\, n]]]\\ [\mathrm{Pr}\, 7]\end{array}}{[\mathrm{D2N}\, 7]}$$

Derivation of the sequent $\mathcal{M}_1, \mathcal{M}_2 \Rightarrow [\mathrm{D2N}\, 7] \cong \mathtt{T}$, where $\mathcal{M}_1 := [\forall^{\nu}\lambda n[[\mathrm{Pr}\, n] \rightarrow [\mathrm{D2N}\, n]]] \cong \mathtt{T}$, $\mathcal{M}_2 := [\mathrm{Pr}\, 7] \cong \mathtt{T}$:

$$\dfrac{\dfrac{\dfrac{\dfrac{\dfrac{}{\mathcal{M}_1 \Rightarrow [\forall^{\nu}[\lambda n[[\mathrm{Pr}\, n] \rightarrow [\mathrm{D2N}\, n]]]] \cong \mathtt{T}}\,(\mathrm{AX})}{\mathcal{M}_1 \Rightarrow [[\lambda n[[\mathrm{Pr}\, n] \rightarrow [\mathrm{D2N}\, x]]]7] \cong \mathtt{T}}\,(\forall\text{-E})}{\mathcal{M}_1 \Rightarrow [[\mathrm{Pr}\, 7] \rightarrow [\mathrm{D2N}\, 7]] \cong \mathtt{T}}\,(\beta\text{-CON})}{\mathcal{M}_1, \mathcal{M}_2 \Rightarrow [[\mathrm{Pr}\, 7] \rightarrow [\mathrm{D2N}\, 7]] \cong \mathtt{T}}\,(\mathrm{WR}) \qquad \dfrac{\dfrac{}{\mathcal{M}_2 \Rightarrow [\mathrm{Pr}\, 7] \cong \mathtt{T}}\,(\mathrm{AX})}{\mathcal{M}_1, \mathcal{M}_2 \Rightarrow [\mathrm{Pr}\, 7] \cong \mathtt{T}}\,(\mathrm{WR})}{\mathcal{M}_1, \mathcal{M}_2 \Rightarrow [\mathrm{D2N}\, 7] \cong \mathtt{T}}\,(\rightarrow\text{-E})$$

3.4 Substitution

In this section, I begin with describing the notion of substitution in the context of λ-calculus (3.4.1) to properly understand its significance for our investigations. I will then define the substitution function $\mathrm{S{\scriptsize UB}}^k$ (3.4.2) which can be employed in constructions that explicitly carry out substitution (3.4.3). Such constructions occur in several rules, notably in the Rule of Substitutivity of Identicals (SI) (3.4.4).

3.4.1 Substitution and conversions in λ-calculus

Recall that λ-calculus was introduced by Church [70] under the name "the *calculus of λ-conversion*". It has three conversion rules, denoted I–III, but they are now called "α-conversion", "β-contraction" (or "β-reduction") and "β-expansion", respectively. The rules are sometimes supplemented by the rules of η-conversion. (Since I write about λ-terms of λ-calculus in this section, I will deliberately use a notation different from that in the rest of the book.)

Every *conversion* enables us to *rewrite* term M to term N while preserving their equivalence (in the sense of identity of denotational semantic values). Thus, for each conversion there exists the smallest equivalence relation such that the couples involved in it are determined by the conversion rule in question. Let "$M \longrightarrow_\gamma N$", where γ equals α, β or η, stands for *γ-contraction* of M to N, while "$N \longrightarrow_\gamma M$" stands for its reverse, i.e. *γ-expansion.* Finite or infinite series of contractions are called *reductions.*[9]

Every conversion can be viewed as a derivation rule based on substitution. Let "$[y/x]M$" denotes the result of substitution of the variable y for all free occurrences of the variable x in term M (in literature, "$M[y/x]$", or "$M[x := y]$" and also "$M\langle x := y\rangle$" or "$M_{(y/x)}$" are used). Here are classical rules of α-, β- and *η-contraction*:

$\lambda x.M$	$\longrightarrow_\alpha$	$\lambda y.[y/x]M$	condition: y does not occur freely in M
$((\lambda x M)\, N)$	$\longrightarrow_\beta$	$[N/x]M$	condition: x does not occur freely in N
$\lambda x(Mx)$	$\longrightarrow_\eta$	$[x/x]M$	condition: x does not occur freely in M

β-contraction is the most important 'computational rule'. It expresses the application of 'function' $\lambda x.M$ to the argument N. α-conversion is based on the fact that the (correct) *renaming of λ-bound variables* does not change the semantics of the term in question. By α-contraction one means the renaming to the lexicographically first variables. η-contraction expresses the Principle of Extensionality of Functions, i.e. the match of 'function' $\lambda x(Mx)$ with 'function' M with regards to their values (given any particular argument).[10]

It is well known that contractions do not preserve equivalence if the substituted variables are inconveniently captured by λ-operators. For example, $[x/y]\lambda x\,(xy)$ leading to $\lambda x\,(xx)$ would not preserve equivalence of the input term $\lambda x\,(xy)$ with the output term. In the case of α-conversion, the right substitution is called the *collisionless renaming* of a variable. For example, $\lambda x\,(xy)$ is rewritten to $\lambda z(zy)$, not to $\lambda y\,(yy)$. With regards to β-contraction, one should prepare the input term by appropriately renaming its λ-bound variables. In our example with $[x/y]\lambda x\,(xy)$, $\lambda x\,(xy)$ should be first converted to $\lambda z\,(zy)$, so the substituted x will not become bound in it.

---

9 The most important ones are *β-reductions.* The calculi involving it, esp. λ_β and $\lambda_{\beta\eta}$, are examined with regards to having the *Church-Rosser property*, which is often called *confluence.* Two terms are confluent iff their β-reductions lead to the term of the same form. When a term is not further reducible, it is called a *term in normal form* or *normalised form.* Typed λ-calculus is strongly *normalisable*, since all its terms can be normalised. Untyped λ-calculus is not strongly normalisable because e.g. $((\lambda x.xx)(\lambda x.xx))$ is a term that cycles, i.e. its β-reduction never ends (the term has no normal form).

10 This is perhaps more visible if η-contraction is defined as $\lambda x(M\,x) \longrightarrow_\eta M$ (condition: x does not occur freely in M).

The substitution that aptly pre-prepares terms to which one substitutes something was defined by Curry and Feys [90]. Curry also removed substitution from extra-theoretic considerations and placed it inside formal systems.

3.4.2 Definition of the substitution function

Tichý [415, 416] offered an adaptation of Curry's and Feys' [90] definition of substitution, which he later extended for his TTT [422]. My definition is based on the definition proposed in Raclavský, Pezlar and Kuchyňka [339], which is an adjustment of Tichý's definition because of our different definitions of the notions of free variable and subconstruction. Moreover, my definition does not define a metalinguistic substitution function (as in Tichý [422]), but its type-theoretically specific case – from which, and two its variants, the metalinguistic definition can be reconstructed.

First, I will define *substitution function* SUB^k which is of type $(*_k *_k *_k *_k)$, for $1 \leqslant k$. It maps ternary strings ('triples') of kth-order constructions D, x, C (where x is a variable) to kth-order constructions which I will denote by "$C_{(D/x)}$" (the scope of the 'operator' "(D/x)" in "$C_{(D/x)}$" is "C").

Notational convention 28 **('$C_{(D/x)}$')**

"$C_{(D/x)}$" stands for a kth-order construction C in which all direct (i.e. kth-order) and free occurrences of the variable x which are free for D are substituted by a kth-order construction D. In other words, $C_{(D/x)}$ is the result of substitution of D for x in C.

Definition 59 (Substitution function, $\mathbf{SUB}^k$)

Let C, D, x, B are kth-order constructions. Let "$FV(C)$" stand for the set of all free variables that are direct subconstructions of C.

I. If the variable x is not free in C, then $C_{(D/x)}$ is identical with C.
II. If the variable x is free in C, then

	If C *is ...*	$C_{(D/x)}$ *is ...*	*condition:*
i.	x	D	
ii.	$[BB_1...B_m]$	$[B_{(D/x)}\ B_{1(D/x)}...B_{m(D/x)}]$	
iii.	$[\lambda y.B]$	$[\lambda y.B_{(D/x)}]$	$x \notin FV(D)$
iv.	$[\lambda y.B]$	$[\lambda z[B_{(z/y)}]]_{(D/x)}$	$y \in FV(D), x \in FV(B)$, $z \notin FV(D) \cup FV(B)$

Remarks. Part I. covers two cases: (i) C does not contain a free variable x, (ii) C does not contain x as its direct (i.e. kth-order) subconstruction. Thus, x is not substitutable e.g. in the (schematic) constructions such as (i) $[\lambda x[...x...]]$ or (ii)

$^0[...x...]$. Substituting only for direct occurrences of x, cf. case (ii), is an important feature. Consider e.g. the sentence such as "$x =^\nu y$ and agent A believes that $[[[1+x] =^\nu 3] = [[1+1] =^\nu x]]$.". If one substitutes something for x in this sentence, she typically wants to substitute something within the left conjunct, not the right one. The occurrence of x in the right conjunct occurs in the scope of the hyperintensional operator of belief (see 5.4.4), so the construction written in the sentence occurs within its logical analysis in the scope of 0-execution. This is why this x (which has a lower order than the other one, it is not a direct subconstruction) is not substitutable.

Example 18 (Substitution function)

Let $w, x, y, z, ^{(0)}3/\nu; ^{(0)}+/(\nu\nu\nu)$.

D	C	$C_{(D/x)}$
y	x	y
y	z	z
$^{(0)}3$	x	^{03}
y	^0x	^0x
$[z+\ w]$	$[x\ +\ y]$	$[[z+w]+y]$
$[z+w]$	$[x+x]$	$[[z+w]+[z+w]]$
$[z+w]$	$[\lambda y[x+y]]$	$[\lambda y[[z+w]+y]]$
$[y+x]$	$[\lambda y[x+y]]$	$[\lambda z[[y+x]+z]]$

Since a variable x may occur freely in ^1E or ^2E (when x is free in E), there are mates of the above substitution function which cover these cases. They are of types $(*_k *_{k-1} *_{k-1} *_k)$ and $(*_k *_{(k-2-l)} *_{(k-2-l)} *_k)$ (l is specified below).

Definition 60 (Substitution function – supplements)

Let C be a construction of order k and E, D, x be constructions of order $k-1$; vii.-case: F, D, x be of order $(k-2-l)$, where l is the number of occurrences of "1" in "(1...1)", for $1 \leqslant (k-2-l)$.

	If C is ...	*$C_{(D/x)}$ is ...*	*condition:*
v.	^1E	$^1E_{(D/x)}$	
vi.	^2E	$^2E_{(D/x)}$	E is not of form $^{(1...1)0}F$
vii.	^2E	$^2[^{(1...1)0}F_{(D/x)}]$	E is of form $^{(1...1)0}F$

In TTT, one can use denotational semantics of $\mathcal{L}_{\mathsf{TTT}}$'s terms for easy control of the fact that the result of the substitution of (say) $[x+2]$ for y in $[y+4]$ $v(1/x;3/y)$-constructs the same object as is $v(1/x;3/y)$-constructed by $[y+4]$. Tichý formulated the general form of such a correlation as the *Compensation Principle* and proved it for both his STT [415] and TTT [422], though one finds there only its version for order 1. In Appendix A, I prove its generalised version:[11]

Theorem 1 (The Compensation Principle)

Let C be any construction of order n. For any valuation v and any construction D of order n, if D v-constructs D, then $C_{(D/x)}$ v-constructs C iff C $v(\mathrm{D}/x)$-constructs C.

The Compensation Principle is a powerful theorem: it enables us to prove rules in which particular clauses of the definition of the substitution function are deployed. Recall also that $C_{(D/x)}$ is short for ${}^2[\mathbf{Sub}^k\,{}^0D\,{}^0x\,{}^0C]$ which means that ${}^0\mathbf{Sub}^k$, which is used in our derivations, is not a metalinguistic device but a genuine operator of our logic. Now I add two general rules in which $C_{(D/x)}$ is used (their proof is trivial given the above theorem).

Definition 61 (Substitution rules)

$$(\mathit{Sub}\text{-I})\ \frac{\Gamma \Rightarrow D \cong x \qquad \Gamma \Rightarrow C \cong a}{\Gamma \Rightarrow C_{(D/x)} \cong a}$$

$$(\mathit{Sub}\text{-E})\ \frac{\Gamma \Rightarrow D \cong x \qquad \Gamma \Rightarrow C_{(D/x)} \cong a}{\Gamma \Rightarrow C \cong a}$$

Condition (*Sub*-I, *Sub*-E): a is not free in C, D, Γ. $D, x/\tau; a, C/\tau_1$. x in C is free for D.

Remark. Generalisation for the case with "$(\bar{D}_m/\bar{x}_m)$" is straightforward.

3.4.3 Explicit substitution

Sub^k is an internal object of TTT and so is v-constructed by various constructions (not only by ${}^{(0)}\mathbf{Sub}^k/(*_k\ *_k\ *_k *_k)$), which can be subconstructions of other constructions. If a construction of Sub^k is composed with a suitable string ('triple') of constructions, the whole composition v-constructs the value of Sub^k in conformity with the above definition.

Remarks. Since explicit substitution inside a logical system is still an extraneous idea in the current paradigm, Abadi, Cardelli, Curien and Levy [1] introduced the

[11] Added in proofs. Duží, Jespersen and Materna [110] published a proof of the generalised Compensation Principle. Two thirds of their proof is in fact Tichý's. The rest is very difficult to follow due to its atypical construction and the fact that according to their definition of rank, constructions of the forms ${}^0C, {}^1C, {}^2C$ (where C is a construction) have each two distinct ranks.

special term "*explicit substitution*", which is now wide-spread. Note, however, that, with regards to TTT, the lowest order of constructions involving $\textsc{Sub}^k$, e.g. $^{(0)}\mathbf{Sub}^k$, is $k{+}1$ and so such constructions are never a part of 'object language' in the sense of kth-order constructions, but of its $(k+1)$st-order 'metalanguage'. It is not clear whether the current approaches to substitution imposes similar restrictions. Anyway, TTT is capable of preventing any paradox resulting from careless confusion of language levels.

Now I am going to show some useful examples. Let $x, {}^{(0)}1, {}^{(0)}2/\nu; {}^0x, {}^{00}1, {}^0[x+2]/*_k$. I will use the symbol "$\underset{v}{\leadsto}$" (introduced in 2.6.3) to indicate the result of v-constructing of the whole construction.

Example 19 (Simple explicit substitution)

$[\mathbf{Sub}^k\ {}^{00}1\ {}^0x\ {}^0[x+2]] \qquad \underset{v}{\leadsto} [1+2]$

Remarks. The construction v-constructs the value of $\textsc{Sub}^k$ for the string ('triple') of constructions $^{(0)}1, x, [x+2]$ which all are introduced in the whole construction by means of 0-execution.

Now let us call D a *substitute.* Except for its 0-execution, i.e. ^0D, there are also other constructions of D. Tichý [422] showed that in such cases the function $\textsc{Triv}^k$ that maps the v-constructed D to its 0-execution (which is unique) is particularly useful (where $^{(0)}\mathbf{Triv}^{(*_k\nu)}/(*_k\nu))$.

Example 20 (Explicit substitution of the received substitute)

$[{}^0\mathbf{Sub}^k\ [\mathbf{Triv}^{(*_k\nu)}\ [3\div y]]\ {}^0x\ {}^0[x+2]] \qquad \underset{v}{\leadsto} [1+2]$, if $v(y)=3$

Both the substitute and the construction to which one substitutes (but even the variable for which one substitutes) can also be obtained by a *nested substitution*:

Example 21 (Nested explicit substitution)

$[\mathbf{Sub}^k\ {}^{00}2\ {}^0y\ [\mathbf{Sub}^k\ {}^{00}1\ {}^0x\ {}^0[x+y]]] \qquad \underset{v}{\leadsto} [1+2]$

In many practical cases, we want to make the result of an explicit substitution, i.e. $C_{(D/x)}$, to v-construct an object (if any); for that sake, a 2-execution will be deployed:

Example 22 (Execution of the result of explicit substitution)

$^2[\mathbf{Sub}^k\ {}^{00}1\ {}^0x\ {}^0[x+2]] \qquad \underset{v}{\leadsto}$ the number 3

Remarks. The whole construction is a $(k+2)$nd-order one (its lowest possible order is 3), and so it comes from 'metametalanguage' with respect to the 'language' containing $[1+2]$. Note also that the difference between $^2[\mathbf{Sub}^k\ {}^{00}1\ {}^0x\ {}^0[x+2]]$ and $[{}^0\mathbf{Sub}^k\ {}^{00}1\ {}^0x\ {}^0[x+2]]$ as such is not visible in the common notation of λ-calculus.

3.4.4 The Rule of Substitutivity of Identicals

I will define the *Rule of Substitutivity of Identicals* (SI) which is an adjustement of Tichý's [415, 416] rule (SI), derived by him from some other rules of his system.

Definition 62 (The Rule of Substitutivity of Identicals, (SI))

$$(\mathrm{SI})\ \frac{\Gamma \Rightarrow {}^2[\mathbf{Sub}^k\ {}^0D_1\ {}^0x\ {}^0C] \cong \mathrm{T} \qquad \Gamma \Rightarrow [D_1 =^{\tau} D_2] \cong \mathrm{T}}{\Gamma \Rightarrow {}^2[\mathbf{Sub}^k\ {}^0D_2\ {}^0x\ {}^0C] \cong \mathrm{T}}$$

Condition: $D_1, D_2, x/\tau; C/o$; x in C is free for D.

For particular examples using (SI), see 5.2, 5.2.3, 5.4.3.

In literature on λ-calculus, we often meet rules capturing the so-called *compatibility* of substitution, according to which substitution within a term may be performed even if the term is a subterm of some greater term. Tichý [415, 416] derived similar rules from the above ones. He also proved Theorem about α-conversion, which I do not repeat here, though I will assume the two relevant rules:

Definition 63 (The Rules of α-Conversion)

$$(\alpha\text{-CON})\ \frac{\Gamma \Rightarrow [\lambda x.{}^2[\mathbf{Sub}^k\ {}^0y\ {}^0x\ {}^0C]] \cong f}{\Gamma \Rightarrow [\lambda x.C] \cong f}$$

$$(\alpha\text{-EXP})\ \frac{\Gamma \Rightarrow [\lambda x.C] \cong f}{\Gamma \Rightarrow [\lambda y.{}^2[\mathbf{Sub}^k\ {}^0y\ {}^0x\ {}^0C]] \cong f}$$

Condition: f is not free in C, Γ. $x, y/\tau; f/(\tau_1\tau); C/\tau_1$. x in C is free for y, and if y is free in C, then y is x.

Finally, I offer *The Rule of Existential Generalisation* (EG), which is an adaptation of Tichý's [416] rule.

Definition 64 (The Rule of Existential Generalisation (EG))

$$(\mathrm{EG})\ \frac{\Gamma \Rightarrow {}^2[\mathbf{Sub}^k\ {}^0D\ {}^0x\ {}^0C] \cong \mathrm{T}}{\Gamma \Rightarrow [\exists^{\tau}[\lambda x.C]] \cong \mathrm{T}}$$

Condition. $x, D/\tau; C/o, {}^{(0)}\exists^{\tau}/(o(o\tau))$. x in C is free for D.

Chapter 4

Transparent hyperintensional logic

4.1 Overview of the chapter

The *semantic doctrine* of *Transparent hyperintensional logic* (THL) is a collection of views on how to explicate meanings of natural language expressions. Many of the views are shared with other approaches,[1] and so I shall only highlight some of the most important ones.

In section 4.2, I expose the neo-Fregean *semantic scheme* of THL, borrowed from Tichý's *Transparent intensional logic* (TIL), according to which the meaning of an expression E is identified with construction C of E's denotatum D. Various famous semantic principles, e.g. that of Compositionality, are in force in THL (4.2.1). Since THL inherits some methods of *possible world semantics* (*PWS*), the system of functions used for the explication of expressions' denotata is based in a type base that involves, *inter alia*, possible worlds (4.3).

(Possible world) *intensions* can be defined in THL as functions from moments of time and possible worlds (4.4). When 'reshuffling' arguments of some intensions, one obtains entities that I shall call *conditions* (4.4.2); conditions are certain functions that have propositions as values. Propositions are nothing but medadic (i.e. 0-ary) conditions. In THL, conditions are used for the explication of the notion of property and an m-ary relation(-in-intension) (for $2 \leqslant m$).[2] The employment of conditions allows for great simplification of THL notation in comparison with TIL.

In section 4.5, I offer an analysis of a fragment of natural language (4.5.1). Then, I propose arguments in favour of THL in comparison with Tichý's TIL

[1] Cf. e.g. Chierchia and McConnell-Ginet [63], Heusinger, Maienborn and Portner [172], Partee, ter Muelen and Wall [288], Lappin [228], Benthem and ter Muelen [36], Aloni and Dekker [5].

[2] For different conceptions of properties, see e.g. Bealer and Mönnich [31].

(4.6). As mentioned e.g. in Preface, the THL system has been proposed by my colleague Kuchyňka in his unpublished manuscripts in 2016. His ideas occur mainly in 4.4.2 and 4.6, partly in 4.4.1, 4.4.3 and 4.5.1.[3]

4.2 Logical analysis and semantic scheme

In accordance with THL (and TIL), the aim of *logical analysis of natural language* is to match an analysed expression E with one – or, in the case of its ambivalence, more than one – construction C, which is proposed as a rigorous explicatum (in Carnap's [57] sense) of its intuitive meaning M.

Definition 65 (Meanings)

Meanings are explicated as constructions.

The choice of explicata, which are often called *logical analyses* or *formalisations*, is governed both by (competent) linguistic intuition and demands of logic. Meanings are understood as abstract objects 'coded' by expressions.

Construction C, proposed as E's logical analysis, is depicted by a term belonging to an extension of the formal language introduced above. One may perhaps speak with Montague [258] about the translation of E into formal language (the idea of translation is contestable, see Raclavský [326]).

As mentioned in chapter 1, since both the TIL and THL approaches are *neo-Fregean*, their *semantic scheme* looks as follows:

Definition 66 (THL semantic scheme)

Expression $E \longrightarrow$ *construction* C (i.e., *meaning*) $\longrightarrow$ *denotatum* D.

Denotatum of E is identified with the object v-constructed by C. So the scheme matches Church's [67, 72] Fregean [143] scheme: E expresses the concept C which determines the denotatum D of that E.

However, in some cases,

$$E \longrightarrow C$$

is all that we have. Consider, e.g. "$3 \div 0$" which has no conceivable reference or denotatum. This particular example might motivate us to reduce the three-membered semantic scheme to the two-membered one, and agree with Tichý [422] that C is all that we need to consider for the explication of E's semantics. Nevertheless, it sometimes makes good sense to also work with denotata within semantic theory.

---

[3] I discussed Kuchyňka's proposal with him. He commented on versions of my papers [322, 299], both written in 2016–17, in which THL had been implemented for the first time.

The construction C offered as an explication of E's meaning v-constructs E's denotatum, which is an object whose type is materially adequate. The association of expressions with denotata of particular types conforms to principles which have been studied e.g. in categorial grammar or compatible systems of linguistic syntactic analysis, which are only implicitly assumed in this book. The semantic doctrine of THL also agrees with some general principles of PWS and, so, some denotata are possible world intensions.

4.2.1 Some semantic principles

Now I am going to briefly discuss four important semantic principles adopted in THL.

i. Construction C, offered as an explication of E's meaning, complies with the *Principle of Compositionality of Meaning*: the meanings of all E's semantically self-contained subexpressions are integrated in E's meaning in such a way that it largely corresponds to E's structure. In ideal cases, an isomorphism occurs between E and C. Divergences from the isomorphism are explained as caused by ellipsis, idioms, etc., i.e. by different coding mechanisms of natural and formal languages.

ii. Construction C, offered as an explication of E's meaning, also meets Carnap's [57] *Principle of Subject Matter* (or *Aboutness*, see e.g. Osorio-Kupferblum [282]). In my words (Raclavský [320]): a (disambiguated) expression E speaks about all, and only those, constructions $C_1, C_2, ...$ that are expressed by E's subexpressions $E_1, E_2, ...$ This principle entails that E's logical analysis should not contain subconstructions that are explicata of expressions which are not present in E.

Remark. My version of the principle relies on constructions expressed by expressions, not on their denotata (which was defended by Duží, Jespersen and Materna [110]). This stance is supported by the following evidence: a competent speaker (see Zouhar [455]) does understand expressions such as "$3 \div 0$.", i.e. grasps their meanings, though with no idea what a possible denotatum is. For another example, consider "natural numbers", the denotatum of which is an infinite object no human can actually grasp in entirety, yet a competent speaker does understand such expressions.

iii. The next principle I mention might be called the *Principle of Harmony of Verification and Meaning*: the meaning of a sentence determines its verification, and *vice versa*. (Verification is considered here as an ideal procedure that need not always be completed by us as finite beings.) A similar principle was the subject of lively discussion by the members of Wiener Kreis and their sympathisers, but perhaps because of the criticism of their theory of science, their 'criterion of meaning' has somewhat been forgotten (with exception of Kaplan [204]). Intuitionistic semantics, see e.g. Martin-Löf [247], revived the principle in a form very close to the one I accept: a sentence such as "Fido is a dog." is seen as a record of finding – alternatively, as a hypothesis, a *problem*[4] – that Fido is a dog and so its verification consists in the execution of a test on Fido's satisfaction of the condition BE A DOG; to put it a bit differently, the proof of the sentence "Fido is a dog." consists in demonstrating that Fido is a dog.

iv. The choice of C as an explication of E's meaning is also affected by an attempt to model inferences that can be made with E in various linguistic contexts. This *Principle of Inferential Consequences Preservation* (as one may perhaps call it) is arguably one of the most fundamental principles of semantic analysis. It enables us to determine an expression's meaning even if one only has rather foggy intuitions about it.

Remarks. To illustrate, the verb "seek" in "Xenia seeks the Fountain of Youth." cannot express a construction of a relation(-in-intension) between individuals, for the sentence does not entail "There is an individual who is the Fountain of Youth.", which would be allowed in such an initial hypothesis. If one adopts the idea that "seek" expresses a construction of a relation(-in-intensions) between individuals and (say) individual 'concepts' that are modelled in terms of possible worlds (see 5.2.2 for discussion), this new hypothesis is tested again (now successfully) by inspecting of linguistic contexts in which the verb possibly occurs. For another example, if a word such as "white" is removed from a linguistic context, one is unsure what it means exactly: in contexts a. "Xenia is white.", b. "white fox", c. "white is her favourite colour", the word "white" stands for a. a property ascribed to an individual, b. a modifier as a function which maps the property BE A FOX to a property with a narrower extension, c. a property which is ascribed some higher-order property. Comparable examples seem to motivate Frege [140] in the formulation of his famous *Context Principle*, which cannot be disapproved given the linguistic evidence.

An obvious next step for somebody who appreciates the Principle of Inferential Consequences Preservation is to adopt some form of *inferentialism*, see e.g. Brandom [47], and to maintain at least some ideas of *proof-theoretic semantics*, see e.g. Schroeder-Heister [364] and Kahle and Schroeder-Heister [199], Sundholm [395] (see below for further references), according to which meanings can be looked upon in terms of principal inferences with their expressions.

The starting point of proof-theoretic semantic was Gentzen's [152] observation concerning proofs and their possible reduction by normalisation, which enabled fixation of the logical idea, also formulated by Gentzen, that introduction rules for logical connectives define them. This topic has been studied e.g. in the important works by Prawitz [305], the volume edited by Piecha and Schroeder-Heister [300], Došen [96], Wieckowski [448], Tennant [403], Tranchi-

[4] For a discussion of the algorithmic approach to problems I accept, see Pezlar [294].

ni [427], Wansing [443], and, using intuitionistic framework, in works by Martin-Löf [245], Dummett [101] and many others.[5]

The first systematic book applying proof-theoretic semantics to natural language expressions instead of logical constants, namely Ranta [342], also descended from an intuitionistic background. For recent works see e.g. Francez and Dyckhoff [136], Francez [135]. As mentioned in the introduction, just this *type-theoretic semantics* marks a progresive line in contemporary formal semantics of natural language, see e.g. Chatzikyriakidis and Luo [62], which Tichý [419] anticipated.

4.3 Type base of **THL**

Proponents of PWS (e.g. Stalnaker [388]), repeatedly stressed that an adequate model of the semantic features of many natural language expressions, one must take into account *modal* and/or *temporal variability*: the fact that the reference of expressions such as "(be a) dog", "the height of Etna", "the president of the USA", "It rains in Paris.", "now" varies depending on *circumstances*, the states of our world and time. In PWS, circumstances are usually understood in terms of *possible worlds* and *moments of time*. The variability in reference is then captured by means of functional abstraction from circumstances. In Tichýan PWS, then, *denotata* of such 'contingent expressions' are explicated by total or partial functions from possible worlds and moments of time, i.e. as *(possible world) intensions*.

The semantic doctrine of THL follows Tichý's TIL, and so for logical analysis of natural language THL, utilises an instance of TTT that has B_{THL} as its base – call it $\mathsf{TTT}_{B_{\mathsf{THL}}}$. The base involves types of possible worlds and moments of time, as described in the next section.

4.3.1 Epistemic framework and **THL**'s base

Every (natural) language is understood to speak about a certain area. For an illustration consider a language $\mathcal{L}$ used in the investigation of the empirical world, which is thus capable of referring to certain objects as individuals, who form the *universe of discourse* U of $\mathcal{L}$, and also to properties, relations and further *attributes*. Of course, $\mathcal{L}$ is also capable of speaking about the actual or merely possible instantiation of these attributes by the members of U (and not only them).

[5] Pezlar [297] shows that proof-theoretic semantics meets hyperintensionality. This can be easily confirmed in my analysis of advanced deductions (see 3.2), for certain inference pieces must occur in the scope of 0-execution.

Some of these attributes can be defined with the help of others; yet not every attribute can be defined, the definitional regress must stop somewhere. Certain attributes are thus understood as being antecedently given and undefined. The collection of such basic attributes is called the *intensional base* IB. U and IB constitute the *epistemic framework* EF of $\mathcal{L}$ in question.[6]

The enterprise of logical analysis of (natural) $\mathcal{L}$ is based on the idealized assumption that EF of the analysed $\mathcal{L}$ is fixed. Then, functions and constructions over a particular *base* B_{THL}, which involves explicata of members of ER, are offered as genuine explicata.

Definition 67 (THL base)

$B_{\mathsf{THL}} ::= \{\iota, o, \omega, \rho, \tau, \upsilon\}$, where

ι	individuals
o	truth values
ω	possible worlds
ρ	real numbers
τ	types
υ	valuations

Remark. B_{THL} contains Tichý's original B_{TIL}, but I add (as in Raclavský, Kuchyňka and Pezlar [339]) types for types and valuations for the sake of analysis of natural language extended by certain mathematical expressions, e.g., the language of this book. The first four members of B_{THL} are discussed in the next four subsections.

Notational convention 29 (Variables ranging over members of B_{THL})

Type	*usual variables ranging over the type*
ι	$x, y, \ldots$
o	$o, o', \ldots$
ω	$w, w', \ldots$
ρ	$t, t', \ldots$

i. Individuals

For the analysis of natural language one usually assumes an infinite number of *individuals*. The individuals in the type ι lack any complexity of material particulars which they explicate, they are logically primitive entities.[7]

[6] In this section, I adopt and sometimes extend a number of Tichý's ideas which appeared (sometimes in a rudimentary form) in his [417, 410, 418, 422], see also Oddie [277].

[7] Tichý [422] defended the metaphysical idea of antiessentialism according to which individuals are bare; see e.g. Raclavský [319] for discussion.

ii. Truth values

By *truth values*, the members of o, I will understand the (abstract) logical objects *True* and *False*, which will be written as

"T" and "F",

respectively. Being treated as primitive, truth values are unanalyzable as well as undefinable. Further truth values are not accepted in standard THL (or TIL), so the framework is still two-valued despite of admission of *truth-values gaps*.

Truth values are often understood as somehow representing the properties of truth/falsity, or of positive/negative knowledge. But Tichý [422] proposed the view that the truth values T and F serve as explicata of the intuitive logical qualities of agreement (Yes) and disagreement (No) which play a part in our intuitive notion of selection.[8]

iii. Real numbers

The members of the type ρ are *real numbers*. Note that they are thus understood as logically primitive. The members of ρ also serve for explication of the moments of time. When doing so, it is standard to represent them by real numbers within the interval $[0, 1]$ in one-to-one fashion, provided, of course, both the origin of the scale and a unit of duration are given (Tichý [422]).

iv. Possible worlds

The members of the atomic type ω are logically primitive objects $W_1, W_2, \ldots, W_n$. The minimal number of worlds in ω is 2.

Since the early 1970s, Tichý (e.g. [417], [410], [422]) maintained that they are surrogates of 'metaphysical' worlds in the sense outlined below (see Raclavský [334] for analysis of Tichý's conception). Yet epistemic understanding of these possible worlds, e.g. in the sense of Hintikka [176], is not in principle excluded and its usage is advisable.

As sketched above, when analysing natural language, I will assume (with Tichý) that every language is underlayed by a range of (intuitive) possible worlds that are generated from combinatorially possible and realizable distributions of attributes from *IB*. To illustrate, consider *IB* only consisting of the attributes BE A WOMAN and BE A MAN.[9] Further consider that the given *EF*'s

[8] This seems to be an original idea among theories of truth values, cf. Shramko and Wansing [373, 372].

[9] Further common kinds of primary attributes include: relations(-in-intension) between individuals, magnitudes, 'propositional attitudes' and also connections ("nexuses") between

U only consists of Xenia and Yannis, symbolised in our example below by "X" and "Y".

All combinatorial possibilities as regards the distribution of attributes from IB through the objects from EF (below, only U) are called *determination systems* of that EF. When ignoring partiality, there are altogether 16 determination systems for EF:

Example 23 (Determination systems)

	BE A WOMAN	BE A MAN
1.	$\{X, Y\}$	$\{X, Y\}$
2.	$\{X, Y\}$	$\{X\}$
3.	$\{X, Y\}$	$\{Y\}$
4.	$\{X, Y\}$	$\emptyset$
5.	$\{X\}$	$\{X, Y\}$
6.	$\{X\}$	$\{X\}$
7.	$\{X\}$	$\{Y\}$
8.	$\{X\}$	$\emptyset$
9.	$\{Y\}$	$\{X, Y\}$
10.	$\{Y\}$	$\{X\}$
11.	$\{Y\}$	$\{Y\}$
12.	$\{Y\}$	$\emptyset$
13.	$\emptyset$	$\{X, Y\}$
14.	$\emptyset$	$\{X\}$
15.	$\emptyset$	$\{Y\}$
16.	$\emptyset$	$\emptyset$

Not every determination system can be considered to be a genuine factual alternative. Attributes are often not mutually independent and thus not every combinatorially possible determination system is realizable. For example, the determination systems in rows 1, 6, 11 each associate one and the same set with the two incompatible attributes. (Admittedly, they can be called, and perhaps even play a role of *impossible possible worlds*, see e.g. Hintikka [175].)

Though every possible world is a determination system, not every determination system is a possible world. In our case, we have 9 possible worlds, namely those in rows 4, 7–8, 10, 12–16. Such worlds are explicated by the logically primitive objects W_1, W_2, ..., W_n, i.e. by the members of ω. It is needless to add that one of the possible worlds represents the current state of affairs, so it happens to have the property BE THE ACTUAL POSSIBLE WORLD.

---

properties or events, e.g., the relation of causality, as argued by Oddie [277] and Oddie with Tichý [425].

Our miniature example only used a finite number of individuals, of primitive attributes, and thus also of possible worlds. But the idea generalizes, and so infinite number of worlds in ω may be considered if the domain of applicability requires it.[10]

4.4 Extensions and intensions in THL

As mentioned above in 2.5.2, various extensional entities such as sets are explicated within TTT by extensions as functions or primitive entities over the base $B_{\{\iota,o\}}$. $B_{\{\iota,o\}}$ is included in B_{THL} and so the explication of those extensional notions is transferred to THL. Now the intuitive entities considered in 4.3 will be identified with intensions as certain functions of $\mathsf{TTT}_{B_{\mathsf{THL}}}$.

4.4.1 Intensions

As mentioned above, PWS defines intensions as functions from possible worlds. Tichý (e.g. [418, 422]) treated them as total or partial functions from world-time couples, $\langle W, T\rangle$s, i.e. as total functions from possible worlds to total or partial functions from moments of time to objects of a given type τ (i.e. a function of type $((\tau\rho)\omega)$). To a certain extent, I depart from Tichý's treatment of intensions. Typically, intensions will be functions from $\langle T, W\rangle$s.

My core motivation is the following: on Tichý's construal, to find out the value of such an intension I in world W and time instant T, one is imagined as beginning with a selection of a particular W and then searching for I's value in a particular T. However, I rather believe that it is much more natural to assume that, for the sake of illustration, an agent asking about how many individuals live on the Earth begins by picking out the current time T (this is the most frequent case) and continues by investigating whether the actual world (at T) belongs to the set of worlds in which there are nine trillion of living creatures, or rather to the set of worlds with eight trillion creatures. Observe that the agent need not find out which particular world is indeed the actual one, for this amounts to becoming omniscient – which is barely attainable. The agent's curiosity will be fully satisfied if she will be sure that the actual world belongs to the set of worlds with (say) nine trillion living individuals.

[10] Possible worlds can alternatively be explicated in a non-trivial way as maximal consistent sets of facts where facts are explicated as certain constructions (see Kuchyňka and Raclavský [221]). Then, a member of ω corresponds to each such set.

Adapting Tichý's definition of intension accordingly leads to the following definition.

Definition 68 (Intension)

A *τ-intension* is a total function from moments of time to total or partial functions from possible worlds to τ-objects, i.e., a function of type $((\tau\omega)\rho)$. Moreover, functions which are equivalent to them, i.e. total functions of type a. $((\tau\rho)\omega)$, or b. $(((\tau_m\omega)\rho)\tau_1...\tau_{m-1})$ (condition: τ of the original τ-intension is of form $(\tau_1, ..., \tau_m)$) which have total or partial functions of type $((\tau_m\omega)\rho)$ as values, will also be called *τ-intensions*.
Instead of τ-intensions, I will usually briefly speak about *intensions*.

Notational convention 30 (Schematic type of intensions, $\tau_{\omega\rho}$)

"$\tau_{\omega\rho}$" is short for "$((\tau\omega)\rho)$".

Remarks. My definition of the notion of intension allows us to call Tichý's intensions "intensions", cf. the a. case. The b. case allows us to call entities I call "conditions" (see 4.4.2) "intensions", too. Moreover, by Schönfinkel's reduction and its inverse, we may also call m-ary functions made from (equivalent) total unary intensions "intensions"; e.g. a function of type $(\iota\rho\omega)$ is an intension, for it is equivalent to an intension of type $((\iota\rho)\omega)$.

I have adjusted Tichý's adaptation of PWS-style explication of some important philosophical entities as follows:[11]

Definition 69 (Important kinds of intensions as explicata)

Entity	*type of the corresponding intension*
office for τ-objects	$\tau_{\omega\rho}$
property of τ-objects	$(o\tau)_{\omega\rho}$
relation(-in-intension) between $\tau_1, ..., \tau_n$-objects	$(o\tau_1...\tau_n)_{\omega\rho}$
proposition	$o_{\omega\rho}$

Notational convention 31 (Type of propositions, π)

"π" is short for "$o_{\omega\rho}$".

[11] See Tichý's [422], [413] (or the translation in [419]) for a philosophical defence of the notion of office. His notion resembles Montague's [258] notion of *individual concept*, but, unlike Montague's (individual) concepts, Tichý's offices can be partial functions; moreover, they are functions from possible worlds and moments of time. For Tichý, any intension is an explicatum of an office; e.g., propositions are offices occupiable by truth values.

In the next section, this explication of properties and relations-in-intensions will be rephrased in terms of conditions.

Both Tichý's and my modified notion of intension allows us to offer Carnap-like [57] parallelism between extensions and intensions:

Example 24 (Frequent types of extensions and intensions)

Extension	*type*	*intension*	*type*
individual	ι	individual office	$\iota_{\omega\rho}$
number	ρ	magnitude	$\rho_{\omega\rho}$
truth value	o	proposition	$o_{\omega\rho}$
set of individuals	$(o\iota)$	property of individuals	$(o\iota)_{\omega\rho}$
function from individuals to numbers	$(\rho\iota)$	function-in-intension from individuals to numbers	$(\rho\iota)_{\omega\rho}$

Example 25 (Extensions and intensions)

Extension	*type*	*intension*	*type*
Xenia	ι	YANNIS' GIRLFRIEND	$\iota_{\omega\rho}$
3.14	ρ	THE HEIGHT OF ETNA (IN METRES)	$\rho_{\omega\rho}$
T	o	THAT IT RAINS IN PARIS	$o_{\omega\rho}$
{Xenia, Yannis}	$(o\iota)$	BE A DOG	$(o\iota)_{\omega\rho}$
$=^{\iota}$	$(o\iota\iota)$	TO LOVE (SOMEBODY)	$(o\iota\iota)_{\omega\rho}$
an arbitrary numbering of individuals	$(\rho\iota)$	THE HEIGHT OF (SOMEBODY) (IN METRES)	$(\rho\iota)_{\omega\rho}$

Having intensions as denotata of certain expressions enables us to separate their denotation from reference. In TIL, the *reference* of such an expression E in W at T is the value of the intension of type $\tau_{\rho\omega}$ denoted by E in W at T. Since THL uses intensions of type $\tau_{\omega\rho}$, or intensions of types equivalent to this type, this feature is also implemented in THL. The reference (in T and W) of expressions which do not denote intensions is identified with their denotation.

Discrimination between meaning, denotation and reference allows us to distinguish semantic relations of synonymy, equivalence and co-reference as follows.[12] *Synonymy* (or *co-expressivity*) amounts to the sameness of meanings in language $\mathcal{L}$; e.g. "catastrophe"–"disaster", "eight"–"8". *Equivalence* (or *co-denotativity*) amounts to the sameness of denotata of expressions in $\mathcal{L}$, which is caused by v-congruence of constructions that are expressed by those expressions; e.g. "the Pope"–"the Archbishop of Rome", "1 + 2"–"$\sqrt{9}$". Unlike equivalence, *co-reference* amounts to the sameness of referents of expressions in $\mathcal{L}$, T and W. In the case of expressions denoting intensions, it is caused by

[12] Here I adopt conception from Raclavský [328], Raclavský, Kuchyňka and Pezlar [339], which completed investigations by Cmorej [78] and Materna [248].

the coextensiion of those intensions in the given T and W; e.g. "the morning star"–"the evening star".

4.4.2 Conditions

In this section, I introduce the notion of *condition*. Since it is not involved in standard PWS approaches, a motivation for this new notion is desirable.

First, I draw on the fact that the notion of condition is already a member of our conceptual armoury. It is natural to say that e.g.,

> the number 2 satisfies the condition [that it is a root of the polynomial] $f(x) = x^2 - 5x + 6$

or that

> Fido satisfies the condition for the World's Ugliest Dog Contest (*WUDC*).

Two important observations should be made now. The second example alludes to a certain condition whose satisfaction depends on a state-of-affairs, for, if fortune had been more favourable to Fido with regards to dog's beauty, Fido would have never qualified for the WUDC. In contrast to this, the mathematical condition is satisfied regardless of the variations of worlds and times; one can thus abstract away from them when considering the condition.

The second observation concerning these conditions is equally important: the same condition can be reported in different ways. The case of the mathematical condition is obvious. Let us, therefore, inspect the non-mathematical one. Assume, for simplicity, that to become a WUDC contestant, the dog must display at least one conspicuous asymmetry in shape or colour; the condition is thus

> (TO) BE A DOG WITH A CONSPICUOUS ASYMMETRY IN SHAPE OR COLOUR,

which will be now abbreviated as "$A(x)$". Fido satisfies $A(x)$, for one of his ears stands up and one flops down in a particularly dreadful manner; call the respective condition "$B(x)$". In addition, Fido has one sinister brown eye paired with a blue faded one, "$C(x)$". Fido is also hairless, "$D(x)$", but some people consider this enthralling. Fido satisfies $A(x)$ regardless of the way one refers to it: "$A(x)$", "$\neg\neg A(x)$", or "$A(x) \vee D(x)$", etc. Moreover, the fact that Fido satisfies $A(x)$ follows from Fido's satisfying $B(x)$; it is similar case with $C(x)$. Thus, non-mathematical conditions are not fine-grained entities, as meanings are; they are structureless, as PWS intensions are.

In addition, they are proposition-like entities: metaphorically speaking, if you combine Fido and the condition (TO) BE A CONTESTANT OF THE WUDC, you obtain the proposition THAT FIDO IS A CONTESTANT OF THE WUDC. While (TO) BE A CONTESTANT OF THE WUDC is a condition satisfiable by particular individuals, which means that the condition is monadic, the condition THAT FIDO IS A CONTESTANT OF THE WUDC is medadic (0-ary). Medadic conditions are best identified with propositions. Of course, conditions satisfiable by m-tuples of entities, for $2 \leqslant m$, also exist.

Definition 70 (Condition)

Conditions are intensions of type $(\pi\tau_1...\tau_m)$, for $1 \leqslant m$, where $\tau_1, ..., \tau_m$ are not necessarily distinct, and are equivalent to intensions of type $(o\tau_1...\tau_m)_{\omega\rho}$.

Arity of a condition	*(schematic) type*	intuitive entity
medadic (or *nulary*)	π	proposition
monadic (or *unary*)	$(\pi\tau_1)$	'property'
dyadic (or *binary*)	$(\pi\tau_1\tau_2)$	'relation-in-intension'
...	...	
polyadic (or m-ary)	$(\pi\tau_1...\tau_m)$	

Example 26 (Conditions)

Example	*type*
THAT IT RAINS IN PARIS	π
BE A DOG	$(\pi\iota)$
TO KICK (SOMEBODY)	$(\pi\iota\iota)$

The core of our proposal thus consists in a shift from Tichý's understanding of (say) properties of individuals as functions of type

$$W \longrightarrow (T \longrightarrow (I \longrightarrow O))$$

to monadic conditions, which are functions of type

$$I \longrightarrow (T \longrightarrow (W \longrightarrow O))$$

Technically, the 'essential step' from TIL to THL thus lies in nothing more than equivalence preserving 'reshuffling' of the argument ranges involved in the function. Below, in 4.6, we will discuss that this 'trick' leads to a considerable saving in notation and has also another benefits.

Philosophically, THL seems to escape the world of PWS. While PWS and TIL accentuate the existence of properties, relations(-in-intension), and propositions, THL can only rely on conditions, though of several sorts. Although properties and relations(-in-intension) can be explicated as, and thus reduced to, conditions, we may keep them in our system of analysis alongside conditions and use them if needed.

4.4.3 Operations on conditions

In this section, I am going to define some frequently used operations on conditions.

I begin with a particularly useful operation called *totalisation of propositions.* It maps any proposition P that is gappy in time instant T and possible world W to the proposition P' which has, in T and W, the value $\mathtt{F}$ instead of the partiality gap (where $o/o; w, w'/\omega; t, t'/\rho$; ${}^{(0)}\exists^{\omega}/(o(o\omega))$; ${}^{(0)} =^{\omega} /(o\omega\omega)$; ${}^{(0)}\mathbf{Tot}^{\pi}/(\pi\pi)$).[13]

Notational convention 32 **('C_{tw}')**

"$[[C\,t]\,w]$", where $w/\omega; t/\rho; C/\tau_{\omega\rho}$, will be abbreviated to "C_{tw}".

Definition 71 **(Totalisation of propositions)**

$$\models [\mathbf{Tot}^{\pi}\,p]_{tw} \Leftrightarrow_o [\exists^{\omega}\lambda w'[p_{tw'} \wedge [w' =^{\omega} w]]]$$

Remark. If $p_{tw'}$ v-constructs nothing (since the value of p is a proposition gappy in the value of w' at given value for t) the whole conjunction is also v-improper and so $[\lambda w'[p_{tw'} \wedge [w' =^{\omega} w]]]$ v-constructs an empty set, which is mapped by $\exists^{\omega}$ to $\mathtt{F}$. If, on the other hand, $p_{tw'}$ is v-proper, $[w' =^{\omega} w]$ plays a role for it checks whether the value of w' matches with that of w.

Now I expose four definitions of notions which link conditions/offices to τ-objects, especially *condition satisfaction* and the (office) *occupancy.* The first notion is my analogue of Tichý's (e.g. [418]) notion of property *instantiation,* while the second notion is essentially Tichý's, it is fundamental for his conception of offices. I supplement the two notions with their weakened forms which I shall call "*comply with*" and "*be*". Let $x/\tau; u/\tau_{\omega\rho}; f/(\pi\tau)$;${}^{(0)}\,\mathbf{Sat}^{\tau}$ (satisfy) $/(\pi\tau(\pi\tau))$;${}^{(0)}\mathbf{Occ}^{\tau}$ (be the occupant of),${}^{(0)}\mathbf{Be}^{\tau}/(\pi\tau\tau_{\omega\rho})$;${}^{(0)}{=}^{\tau}\,/(o\tau\tau)$; ${}^{(0)}\mathbf{CompW}^{\tau}$ (comply with) $/(\pi\tau(\pi\tau))$.

Definition 72 **(Satisfy, comply with, occupy, be)**

$$\models [\mathbf{Sat}^{\tau}\,x\,f]_{tw} \Leftrightarrow_o [\mathbf{Tot}^{\pi}\,[f\,x]]_{tw}$$
$$\models [\mathbf{CompW}^{\tau}\,x\,f]_{tw} \Leftrightarrow_o [f\,x]_{tw}$$
$$\models [\mathbf{Occ}^{\tau}\,x\,u]_{tw} \Leftrightarrow_o [\mathbf{Tot}^{\pi}\,[\lambda t'\lambda w'[x =^{\tau} u_{tw}]]]_{tw}$$
$$\models [\mathbf{Be}^{\tau}\,x\,u]_{tw} \Leftrightarrow_o [x =^{\tau} u_{tw}]$$

Remarks. The notion COMPLY (WITH) can be neither applicable, nor counter-applicable to a couple ⟨τ-object, condition for τ-objects⟩. To illustrate, a certain quark Q is neither a women, nor an entity, which is not a woman (this case should not be confused with being a non-woman, which is a notion applicable to Q). In

---

[13] This Kuchyňka's definition may replace the equivalent definition of the 'strong truth predicate' in Raclavský [323], cf. also the operator Tot in 3.3.3.

contrast to this, the notion SATISFY is applicable, or counter-applicable to couples ⟨τ-object, condition for τ-objects⟩. An analogous couple of claims applies to BE (SOMETHING) and OCCUPY (SOMETHING). The two notions will be employed in 4.5.5 below.

Compression of my analyses will be achieved by utilising the following symbols representing important *operations on conditions.* In the case of operations on propositions, the symbols can be understood as *'intensional connectives'*. To denote the operations, I will write small "π" under familiar symbols such as "$\wedge$", getting thus "$\underset{\pi}{\wedge}$".[14] Let ${}^{(0)}\neg/(oo);{}^{(0)}\wedge,{}^{(0)}\vee,{}^{(0)}\rightarrow/(ooo);p,q,p'/\pi$ – where p' is distinct from p and q; ${}^{(0)}\underset{\pi}{\neg}/(\pi\pi);{}^{(0)}\underset{\pi}{\wedge},{}^{(0)}\underset{\pi}{\vee},{}^{(0)}\underset{\pi}{\rightarrow}/(\pi\pi\pi)$.

Definition 73 (Basic operations on medadic conditions I)

$$\begin{array}{lll}\models [\underset{\pi}{\neg}\, p] & \Leftrightarrow_{p'} & [\lambda t\lambda w[\neg\, p_{tw}]]\\ \models [p \underset{\pi}{\wedge} q] & \Leftrightarrow_{p'} & [\lambda t\lambda w[p_{tw} \wedge q_{tw}]]\\ \models [p \underset{\pi}{\vee} q] & \Leftrightarrow_{p'} & [\lambda t\lambda w[p_{tw} \vee q_{tw}]]\\ \models [p \underset{\pi}{\rightarrow} q] & \Leftrightarrow_{p'} & [\lambda t\lambda w[p_{tw} \rightarrow q_{tw}]]\end{array}$$

It is easy to adopt the above definitions for monadic (dyadic, ...) conditions. For example, the 'conjuction' of monadic properties is definable as (where $f,g,f'/(\pi\tau)$ – variables for monadic conditions of τ-objects, where f' is distinct from f and g; x/τ; ${}^{(0)}\forall^{\tau}/(o(o\tau));{}^{(0)}\Cap/((\pi\tau)(\pi\tau)(\pi\tau))$):

Example 27 ('Conjunctive' monadic condition)

$$\models [f \Cap g] \quad \Leftrightarrow_{f'} [\lambda t\lambda w[\forall^{\tau}\lambda x[[f\,x]_{tw} \wedge [g\,x]_{tw}]]]$$

Finally, it is also sometimes convenient to have 'intensional quantifiers' at our disposal (where ${}^{(0)}\exists^{\tau}/(o(o\tau));{}^{(0)}\underset{\pi}{\exists^{\tau}},{}^{(0)}\underset{\pi}{\forall^{\tau}}/(\pi(\pi\tau))$):

Definition 74 (Basic operations on monadic conditions II)

$$\begin{array}{lll}\models [\underset{\pi}{\forall^{\tau}}\, f]_{tw} & \Leftrightarrow_{o} & [\forall^{\tau}[\lambda x[f\,x]_{tw}]]\\ \models [\underset{\pi}{\exists^{\tau}}\, f]_{tw} & \Leftrightarrow_{o} & [\exists^{\tau}[\lambda x[f\,x]_{tw}]]\end{array}$$

[14] One can imagine an alternative notation in which (say) "$\sim$", "&", "v" and "$\supset$" are reserved for boolean operations, but "$\neg$", "$\wedge$", "$\vee$" and "$\rightarrow$" are reserved for operations on propositions. Cf. Raclavský [322] where the notation is used.

4.5 Constructions as meanings in THL

As mentioned in chapter 1 and also in the first part of this chapter, constructions are convenient for the explication of 'intensional' entities with a fine-grained structure, most notably meanings of language expressions. In the following (sub)sections, I show sample THL analyses of some unproblematic natural language expressions, most of them are comparable to analyses provided by TIL,[15] which I discuss in section 4.6 below. In their various writings, Duží and Jespersen (e.g. [108, 106], [105, 104], [195], with Carrara [196]) developed and applied TIL to various directions and areas, many of them translatable to THL.

Definition 75 (Logical form)

By *logical form* LF of an expression E I mean the construction which is its adequate logical analysis (i.e. E's meaning), in which all 0-executions of extra-logical objects are correctly replaced by type-theoretically appropriate variables.

Remarks. Tichý's unfinished monograph *Meaning Driven Grammar* (cf. Tichý's papers [421, 412]) offers many more complicated units of analysis, so-called semantic pairs, whose semantic components are constructions, while their syntactic components also contain key linguistic characteristics such as gender or an anaphoric parameter. Tichý showed the main derivation rules involving the constructions.

I add a stipulation concerning explication of '(structured) propositions':

Definition 76 ('Propositions' as o-constructions)

Structured, fine-grained *'propositions'* are (explicated by, identified with) constructions of truth values, i.e. o-constructions.

4.5.1 Analysis of singular terms, connectives, predicates

I begin with (a) *proper names*, (b) *connectives*, (c) *'operators'*, (d) simple/compound *predicates* (or *noun phrases*), (e) *modifiers*. For analyses of descriptions and quantifiers see below.

[15] My examples are largely adapted from Raclavský, Kuchyňka and Pezlar [339], many of them are originally Tichý's [422, 419], cf. also Raclavský [328], Duží, Jespersen and Materna [110].

Example 28 (Proper names, connectives, predicates, modifiers)

	Expression	*type of denotatum*	*construction*
(a)	"Xenia"	ι	$^{(0)}\mathbf{X}$
	"zero"; "one"; "two"; "three"	ρ	$^{(0)}0$; $^{(0)}1$; $^{(0)}2$; $^{(0)}3$
(b)	"not"	(oo) / $(\pi\pi)$	$^{(0)}\neg$ / $^{(0)}\underset{\pi}{\neg}$
	"or"	(ooo) / $(\pi\pi\pi)$	$^{(0)}\vee$ / $^{(0)}\underset{\pi}{\vee}$
(c)	"be identical (with)"	$(o\tau\tau)$	$^{(0)} =^{\tau}$
	"plus"	$(\rho\rho\rho)$	$^{(0)}+$
	"the"	$(\tau(o\tau))$	$^{(0)}\imath^{\tau}$
(d)	"be a dog"; "be white"	$(\pi\iota)$	$^{(0)}\mathbf{Dog}$; $^{(0)}\mathbf{Wh}$
	"be prime"	$(o\rho)$	$^{(0)}\mathbf{Pr}$
	"be a dog or white"	$(\pi\iota)$	$[\lambda x[[\mathbf{Dog}\, x] \underset{\pi}{\vee} [\mathbf{Wh}\, x]]$
(e)	"be white [dog]"	$((\pi\iota)(\pi\iota))$	$^{(0)}\mathbf{Wh2}$

Remarks. For reasons of brevity, I provide no discussion on the notorious problems (see e.g. Salmon [359]) related to proper names here. I understood them simply as (unique) 'labels' for individuals (or other objects). Some names are in fact disguised descriptions (see more on this below). In this book, I ignore the difference between predicates and nouns; thus e.g., "dog", "be a dog", "being a dog" are treated alike. The analysis of "the" utilises the *singularisation function* $\imath^{\tau}$ ('iota operator'), see 2.5.1, 3.3.3. The linguistic evidence does not provide a clear difference between *extensional* and *intensional reading* of connectives and other expressions, and so I will usually show my preferred choice, not all possible readings.

4.5.2 Analysis of simple sentences

Here are some unproblematic examples of *sentences* (for the analyses of identity statements, quantified sentences and belief sentences, see below). Every sentence refers (in a given world and time-instant) to a truth value (if any), i.e. an object of type o. Here, I deliberately accentuate the *extensional understanding* of sentences on which they are vehicles of reference to truth values (note that speakers usually do not know to which truth value a sentence refers in a particular world and time-instant). I do not refuse the *intensional understanding* on which they denote propositions, but I keep it for intensional contexts (e.g. "it is necessary that ...", 6.3.1); see the discussion in 4.6 below.[16]

[16] The construction in the last row is β-reducible (3.3.2) to the construction in third row.

Example 29 (Some simple sentences)

Sentence	*construction*
"3 is prime."	$[\mathbf{Pr}\,3]$
"$1+2=3$."	$[[1+2] =^{\rho} 3]$
"Xenia is a dog."	$[\mathbf{Dog}\,\mathbf{X}]_{tw}$
"Xenia is not a dog."	$[\underset{\pi}{\neg}[\mathbf{Dog}\,\mathbf{X}]]_{tw}$
"Xenia is a dog or 3 is prime."	$[[\mathbf{Dog}\,\mathbf{X}]_{tw} \vee [\mathbf{Pr}\,3]]$
"Xenia is a dog or Xenia is not a dog."	$[[\mathbf{Dog}\,\mathbf{X}] \underset{\pi}{\vee} [\underset{\pi}{\neg}[\mathbf{Dog}\,\mathbf{X}]]]_{tw}$
"Xenia is such that she is a dog."	$[[\lambda x[\mathbf{Dog}\,x]]\ \mathbf{X}]_{tw}$

4.5.3 Analysis of descriptions

The present analysis of *descriptions* is largely adopted from Tichý [410, 422].[17] On their *intensional reading*, descriptions denote offices, while the offices are occupiable by certain objects depending on time and world. On their *extensional reading* (not considered by Tichý), descriptions are used to refer to an individual (if any), or an object of another type, which satisfies the description in the time (and world) of its utterance.

The following examples are analysed in their intensional reading; the denotata of all my next examples are of type $\iota_{\omega\rho}$.

Example 30 (Descriptions)

	Description	*construction*
(a)	"[the] Pope"	${}^{(0)}\mathbf{P}$
(b)	"the President of the USA"	$[\lambda t\lambda w[\imath^{\iota}[\lambda x[\mathbf{Pres}\,x\ \mathbf{USA}]_{tw}]]]$
(c)	"Xenia's dog"	$[\lambda t\lambda w[\imath^{\iota}[\lambda x[\mathbf{Dog2}\,x\ \mathbf{X}]_{tw}]]]$
(d)	"the morning star"; "the evening star"	${}^{(0)}\mathbf{MS}$; ${}^{(0)}\mathbf{ES}$
(e)	"Pegasus"	${}^{(0)}\mathbf{Pe}$

Remarks. (b) utilises the dyadic condition BE THE PRESIDENT OF, i.e., ${}^{(0)}\mathbf{Pres}$ (be a president of) $/(\pi\iota\iota)$. Similarly for (c), ${}^{(0)}\mathbf{Dog2}$ (be a dog of) $/(\pi\iota\iota)$. The USA is analysed simply as an individual, ${}^{(0)}\mathbf{USA}/\iota$. To simplify many of our analyses below, I will often treat the whole descriptive phrase as unanalysable (yet denoting an office). Arguably, the descriptions in (d) and (e) are *hidden descriptions* ('descriptions in disguise'), not proper names, since their reference conceivably varies across logical space and time scale. "Pegasus" is a currently non-referring hidden description. It is an example of an '*empty term*', similarly as the *improper description* "the King of France".

[17] See Raclavský [328] for extensive investigation and application.

In his paradigmatic analysis of descriptions, Russell [354] captured major intuitions with regards to descriptions, which is why his analysis was largely accepted. Even today many logicians accept the theory despite two important objections to it, one by Strawson (see 4.5.5), the second by Church [67] (see also Kaplan [204]). Church's objection is in fact fatal: according to Russell's method of description elimination, a sentence such as "Xenia seeks the Fountain of Youth." is formalised as $\exists x(F(x) \wedge S(X,x) \wedge \forall y(F(y) \rightarrow (y = x)))$, from which it follows that the Fountain of Youth exists, $\exists x F(x)$. But this is surely not entailed by the former sentence, for Xenia's search is logically independent on the fountainhead's existence. Tichý's analysis, which I adopt, does not have such an undesirable existential import,[18] being thus in accordance with linguistic intuition (see 5.2.2 for more).

4.5.4 Analysis of identity statements

Identity statements are frequently used in language but many of them are characteristically amenable to *extensional* or *intensional reading* (in a different sense than used above), to which I add *hyperintensional reading*, or their 'crossed' variants. The intended reading of a sentence is often detectable by the exclusion of currently false variants. The referents (if any) of all our examplary sentences are again of type o. Let ${}^{(0)}\mathbf{TC}/\iota; {}^{(0)}\mathbf{Wa}, {}^{(0)}\mathbf{H_2O}/(\pi\iota)$.

Example 31 (Identity statements)

	Sentence	*construction*
(a)	"Tullius is Cicero."	$[\mathbf{TC} =^{\iota} \mathbf{TC}]$
(b)	"$3 \div 0 = 1$."	$[[3 \div 0] =^{\rho} 1]$
(c)	"Xenia is the morning star."	$[\mathbf{X} =^{\iota} \mathbf{MS}_{tw}]$
(d)	"The morning star is the evening star."	$[\mathbf{MS}_{tw} =^{\iota} \mathbf{ES}_{tw}]$
	on its intensional reading:	$[\mathbf{MS} =^{\iota_{\omega\rho}} \mathbf{ES}]$
(e)	"Water is H_2O."	$[[\lambda x[\mathbf{Wa}\, x]_{tw}] =^{(o\iota)} [\lambda x[\mathbf{H_2O}\, x]_{tw}]]$
	on its intensional reading:	$[\mathbf{Wa} =^{(\pi\iota)} \mathbf{H_2O}]$
(f)	"$1 + 2$ is $2 + 1$."	$[[1 + 2] =^{\rho} [2 + 1]]$
	on its hyperintens. reading:	$[{}^{0}[1 + 2] =^{*_1} {}^{0}[2 + 1]]$

Remarks. Sentence (a) expresses the self-identity of Tullius using his two synonymous names. The sentence is informative for those speakers who are not fully competent to understand the current English. (a) was discussed in what is called the *New Frege's Puzzle* (e.g. Salmon [360]), in which the premises of the following argument are valid, but its conclusion is not – contrary to what the direct reference theory of proper names, proposed by Kripke [219] and others, predicts:

[18] See e.g. my analysis of the notion in Raclavský [322].

Example 32 (New Frege's Puzzle)

$$\frac{\text{“Xenia knows that Cicero is Cicero.”} \quad \text{“Tully is Cicero.”}}{\text{“Xenia knows that Tully is Cicero.”}} \text{(SI)}$$

In my view, the alleged falsity of the conclusion and truth of the premises clearly indicates Xenia's insufficient competence to understand English. (The semantic doctrine of e.g. THL is not affected by this, since it abstracts from incompetent speakers.) (d) is factually true, for the two descriptions involved in it are co-referential in the current T and W. On its intensional reading, however, (d) is a priori false, for the two descriptions are not co-denotative. A similar couple of comments applies to (e).[19] *Mutatis mutandis*, it also applies to (f).

4.5.5 Sentences without a truth value

Being framed in TTT, both TIL and THL manage the analysis of expressions related to partiality, esp. sentences that contain improper descriptions. For example, mathematical expressions such as "$3 \div 0$" are analysed as expressing a construction involving the partial function $\div$, and so "$3 \div 0$" is treated as referring to nothing whatsoever.

Consequently, the sentence (formalised in the preceding section)

"$3 \div 0 = 1$."

lacks a truth value (it is *gappy*), for, since there is no number delivered by $3 \div 0$, there is no couple of numbers to which $=^\rho$ would be applicable, or counter-applicable. This phenomenon has clear analogues in some non-mathematical sentences, e.g. "The King of France is bald.", whose gappiness was noticed and defended by Strawson [393, 392].

Now I introduce the idea of *'weak'* and *'strong' reading* of a conceivably gappy sentence. Here is a convenient example (where $^{(0)}\mathbf{KF}$ (the King of France) $/\iota_{\omega\rho}$ – for simplification (see 4.5.1); $^{(0)}\mathbf{Be}^\iota, ^{(0)}\mathbf{Occ}^\iota/(\pi\iota\iota_{\omega\rho})$):

Example 33 ('Weak' and 'strong' reading of identity statement with improper description)

Sentence	*construction*
"Xenia is the King of France."	
'weak' reading:	$[\mathbf{X} =^\iota \mathbf{KF}_{tw}]$
its equivalent:	$[\mathbf{Be}^\iota\, \mathbf{X}\, \mathbf{KF}]_{tw}$
'strong' reading:	$[\mathbf{Occ}^\iota\, \mathbf{X}\, \mathbf{KF}]_{tw}$

[19] There are, of course, possible worlds and moments of time in which English is different from its current stage, and the noun phrases "water" and "H_2O" are co-denotative, or even synonymous in it, and thus in both cases co-referential.

The above-analysed sentence is gappy on its 'weak' reading on which $=^{\iota}$ should be applied to the (in fact missing) referent of "the King of France" (cf. also the examples in the previous subsection). This reading is v-congruent with the analysis employing the notion BE (SOMETHING) (4.4.3).

In the case of a 'strong' reading, however, the sentence is false, because Xenia is definitely not in the relation(-in-intension) towards the office THE KING OF FRANCE. For practical purposes, the negation of this reading is useful, since such a sentence refers to the truth value T, unlike the negation of its 'weak' reading, which refers to no truth value, as its non-negated form.

As mentioned when defining the notions SATISFY and COMPLY (WITH), the former notion is used in the 'strong' reading, while the latter notion is used in the 'weak' reading of sentences in which a relation(-in-intension) towards a certain condition is attributed to something. For example, the sentence

"Xenia is the (only) sister of the King of France."

is false in its 'strong' reading, but it is without a truth value in its 'weak' reading.

If we put an improper description into sentence subject position, an analogue of our 'weak' and 'strong' reading is *extensional* and *intensional reading.* In one of the next analyses, I will use the notion BE A REQUISITE (OF SOMETHING) which I define below (where ${}^{(0)}\mathbf{Ba}$ (be bald)$/(\pi\iota)$;${}^{(0)}$ $\mathbf{Req}^{\iota}/(o(\pi\iota)\iota_{\omega\rho})$):

Example 34 (Extensional and intensional reading of sentence with improper description)

Sentence	*construction*
"The King of France is bald."	
extensional reading:	$[\mathbf{Ba}\,\mathbf{KF}_{tw}]_{tw}$
intensional reading:	$[\mathbf{Req}^{\iota}\,\mathbf{Ba}\,\mathbf{KF}]$

If the description is improper, there is no object to which something could be attributed, the sentence is, on its extensional reading, without a truth value. On the other hand, its intensional reading takes the sentence to be an analytical statement, though a false one.[20]

Here is my adaptation of Tichý's [418, 414] (see Raclavský [328]) definitions of the two types of the notion of *requisite*, which are implicitly assumed in a sentence such as "The King of France is bald." or "Horses are (by definition) mammals." on their intensional reading (where $x/\tau; u/\tau_{\omega\rho}$;

[20] If, on the other hand, the King of France were bald by definition, the statement would be analytically true, as the statement "The President of the USA is a president." is on its intensional reading. There is an obvious relation of requisites to the topic of *meaning postulates*, see e.g. Zimmerman [454].

$f, g/(\pi\tau)$; ${}^{(0)}\mathbf{Req}^{\tau}/(o(\pi\tau)\tau_{\omega\rho})$; ${}^{(0)}\mathbf{Req}^{\mathbf{F}\tau}/\ (o(\pi\tau)(\pi\tau))$; ${}^{(0)}\forall^{\rho}/(o(o\rho))$; ${}^{(0)}\forall^{\omega}/(o(o\omega)))$:

Definition 77 (Requisite of an office/condition)

$$\models [\mathbf{Req}^{\tau}\, f\, u] \ \Leftrightarrow_o [\forall^{\rho}\lambda t[\forall^{\omega}\lambda w[\forall^{\tau}\lambda x[[\mathbf{Occ}^{\tau}\, x\, u] \underset{\pi}{\rightarrow} [\mathbf{Sat}^{\tau}\, x\, f]]_{tw}]]]$$

$$\models [\mathbf{Req}^{\mathbf{F}\tau}\, g\, f] \Leftrightarrow_o [\forall^{\rho}\lambda t[\forall^{\omega}\lambda w[\forall^{\tau}\lambda x[[\mathbf{Sat}^{\tau}\, x\, f] \underset{\pi}{\rightarrow} [\mathbf{Sat}^{\tau}\, x\, g]]_{tw}]]]$$

Moreover, THL is capable of analysing sentences involving a *category mistake*, since TTT allows compositions which may apply a function to a type-theoretically wrong argument (Raclavský [328]). Cf. a slight modification of Carnap's famous example:

Example 35 (Sentence involving category mistake)

Sentence	*construction*
"Xenia is a prime."	$[\mathbf{Pr}\,\mathbf{X}]$

4.5.6 Analysis of expressions with quantifiers

Since 1980s, the topic of *generalised quantifiers* has been rather popular in semantical analysis of natural language.[21] The quantifiers "somebody", "everybody", and also various numeric quantifiers such as "at least three" are understood as denoting classical medadic generalized quantifiers of type $(o(o\tau))$. Unary generalized quantifiers are then of type $((o(o\tau))(o\tau))$, e.g. "some", "all"; binary ones are of type $((o(o\tau))(o\tau)(o\tau))$, e.g. "as much as"; etc.

I begin with definitions of three basic unary generalized quantifiers (where $s_1, s_2/(o\tau)$ – variables for sets; ${}^{(0)}\mathbf{All}^{\tau}$, ${}^{(0)}\mathbf{Some}^{\tau}$, ${}^{(0)}\mathbf{No}^{\tau}/((o(o\tau))(o\tau))$; ${}^{(0)}\mathbf{Tot}^{(o\iota)}/((o\iota)(o\iota))$ – a type-theoretic version of ${}^{(0)}\mathbf{Tot}^{\pi}$, 4.4.3):

Definition 78 (Unary generalized quantifiers)

$$\begin{array}{lll}
\models [[\mathbf{All}^{\tau}\, s_1]\, s_2] & \Leftrightarrow_o & [\forall^{\tau}\lambda x[[\mathbf{Tot}^{(o\tau)}\, [s_1\, x]] \rightarrow [\mathbf{Tot}^{(o\tau)}\, [s_2\, x]]]] \\
\models [[\mathbf{Some}^{\tau}\, s_1]\, s_2] & \Leftrightarrow_o & [\exists^{\tau}\lambda x[[s_1\, x] \wedge [s_2\, x]]] \\
\models [[\mathbf{No}^{\tau}\, s_1]\, s_2] & \Leftrightarrow_o & [\forall^{\tau}\lambda x[[\mathbf{Tot}^{(o\iota)}\, [s_1\, x]] \rightarrow [\neg[\mathbf{Tot}^{(o\iota)}\, [s_2\, x]]]]]
\end{array}$$

In THL, one often utilises their 'intensional' mates (where $f_1, f_2/(\pi\tau)$; ${}^{(0)}\underset{\pi}{\mathbf{All}}^{\tau}$, ${}^{(0)}\underset{\pi}{\mathbf{Some}}^{\tau}$, ${}^{(0)}\underset{\pi}{\mathbf{No}}^{\tau}/((\pi(\pi\tau))(\pi\tau)))$:

[21] For example, Barwise [25], Barwise and Cooper [26], Sher [371], Westerståhl [444], Peters and Westerståhl [293]. In this subsection, I adapt and extend results from Tichý [418], Raclavský [328], Raclavský, Kuchyňka and Pezlar [339]. For TIL and THL investigation of the Square of Opposition, and even in its modal variant, see Raclavský [335, 322].

Definition 79 ('Intensional' unary generalized quantifiers)

$$\models [[\mathbf{All}^{\tau}_{\pi}\, f_1]\, f_2]_{tw} \quad \Leftrightarrow_o \quad [[\mathbf{All}^{\tau}\, [\lambda x[f_1\, x]_{tw}]]\, [\lambda x[f_2\, x]_{tw}]]$$
$$\models [[\mathbf{Some}^{\tau}_{\pi}\, f_1]\, f_2]_{tw} \quad \Leftrightarrow_o \quad [[\mathbf{Some}^{\tau}\, [\lambda x[f_1\, x]_{tw}]]\, [\lambda x[f_2\, x]_{tw}]]$$
$$\models [[\mathbf{No}^{\tau}_{\pi}\, f_1]\, f_2]_{tw} \quad \Leftrightarrow_o \quad [[\mathbf{No}^{\tau}\, [\lambda x[f_1\, x]_{tw}]]\, [\lambda x[f_2\, x]_{tw}]]$$

Here are some examples (where $x_1, x_2/\iota$; ${}^{(0)}\mathbf{Bo}$ (be a boy), ${}^{(0)}\mathbf{Gi}$ (be a girl) / $(\pi\iota)$; ${}^{(0)}\mathbf{Li}$ (to like)/$(\pi\iota\iota)$):

Example 36 (Sentences with quantifiers)

Sentence	*construction*
"No dogs are white."	$[[\mathbf{No}^{\iota}_{\pi}\, \mathbf{Dog}]\, \mathbf{Wh}]_{tw}$
"Some boys like all girls."	$[[\mathbf{Some}^{\iota}_{\pi}\, \mathbf{Bo}][\lambda x_1[[\mathbf{All}^{\iota}_{\pi}\, \mathbf{Gi}][\lambda x_2[\mathbf{Li}\, x_1\, x_2]]]]]_{tw}$

4.6 Comparing THL with TIL

Since THL is relatively new and TIL is an established system, THL's advantages over TIL should be stated. The following list is not exhaustive, nor is it philosophically unassailable (this need not concern us, however, since the choice of THL or TIL is a pragmatic matter anyway).

(i) There is a group of more or less philosophically relevant, but nevertheless somewhat elusive reasons in favour of THL mentioned in 4.4.2 when exposing the notion of condition: a certain metaphysical sobriety of the very notion in comparison to the notions of property and relation(-in-intension), more natural order of steps when evaluating a sentence, and perhaps some others.

Yet the last fact has a significant, if not ultimate, appeal. Observe that in both Montague's and Tichý's understanding (both described in Tichý [424]) of a sentence such as

> "Fido is a dog."

its meaning does not fit the probable method of its verification; i.e., it does not meet the above-mentioned Principle of Harmony of Verification and Meaning (4.2.1). If one would take Montague's or Tichý's analysis as a recipe for verification, in carrying out a verification one would have to i. select a particular world W (and time instant T), ii. find the extension of the property BE A DOG in W (at T), getting thus a certain set, iii. checking whether Fido belongs to it.

Such a method of verification is usually impossible to implement and so it is not used in practice. When, for example, a European biologist checked whether a particular platypus X is a mammal, attempting thus to verify the sentence "X is a mammal.", she hardly checked all individuals on Earth and other planets to find out which set is the extension of the property BE A MAMMAL, and inquired then whether X is one of its member.

In the THL picture, on the other hand, the biologist checked whether X satisfies the condition BE A MAMMAL. Satisfaction of the condition is entailed e.g. by satisfying the condition BREAST-FEED ONE'S OFFSPRING – which is easy to confirm when pursuing X in the Australian bush, there is no need to inspect the whole world. (Cf. also my Fido–WUDC example.)[22]

(ii) One of the most persuasive reasons for the adoption of THL is rather prosaic: the prevailing terseness of THL analyses in comparison with TIL. The following examples enable us to see that crucial tenets of TIL (sentences denote propositions, predicates denote properties/relations(-in-intension), etc.) are responsible for the clumsiness of TIL analyses, since they cause possible worlds and moments of time to be repeatedly employed via variables but then neglected using λ-abstraction.[23]

In TIL analyses, intensions equivalent to intensions utilised in THL analyses will play a part. Let ${}^{(0)}\mathbf{TPrU}$ (the President of USA) $/\iota_{\rho\omega}$ (the simplified TIL analysis) and ${}^{(0)}\mathbf{PrU}$ (the President of USA) $/\iota_{\omega\rho}$ (the simplified THL analysis; see 4.5.1, point iii); ${}^{(0)}\underset{\pi}{\Box}/(\pi\pi)$;${}^{(0)}\,\Box/(o\pi)$ (see 6.3 for more on modalities); ${}^{(0)}\mathbf{TWh}$ (be white) $/(o\iota)_{\rho\omega}$ (a property as analysed by TIL); ${}^{(0)}\mathbf{TBel}^1/(o\iota{*_1})_{\rho\omega}$ (the verb "believe" as analysed by TIL); ${}^{(0)}\mathbf{Bel}^1/(\pi\iota{*_1})$ (the verb "believe" as analysed by THL); cf. below and mainly chapter 5.

Example 37 (Comparison of THL and TIL analyses I)

THL *analysis*	TIL *analysis*
$[\mathbf{Wh}\,\mathbf{X}]_{tw}$	$[\lambda w\lambda t[\mathbf{TWh}_{wt}\,\mathbf{X}]]$
$[[1+2] =^{\rho} 3]$	$[\lambda w\lambda t[[1+2] =^{\rho} 3]]$
$[[\mathbf{Wh}\,\mathbf{X}]_{tw} \vee [[1+2] =^{\rho} 3]]$	$[\lambda w\lambda t[[\lambda w\lambda t[\mathbf{TWh}_{wt}\,\mathbf{X}]]_{wt}$ $\vee[\lambda w\lambda t[[1+2] =^{\rho} 3]]_{wt}]]$
$[\mathbf{Bel}^1\,\mathbf{X}\,{}^0([\mathbf{Wh}\,\mathbf{PrU}_{tw}]_{tw})]_{tw}$	$[\lambda w\lambda t[\mathbf{TBel}^1\,\mathbf{X}$ ${}^0[\lambda w\lambda t[\mathbf{TWh}_{wt}\,\mathbf{TPrU}_{wt}]]]]$
$[\underset{\pi}{\Box}[\mathbf{Wh}\,\mathbf{X}]]_{tw}$	$[\lambda w\lambda t[\Box[\lambda w\lambda t[\mathbf{TWh}_{wt}\,\mathbf{X}]]]]$

Some TIL terms thus translate to THL terms via the *translational rule* $(\cdot)^{\bullet}$ whose essential part for sentences looks as follows. Let $TF/(o\tau)_{\rho\omega}$ (a property

[22] Similar considerations led Barwise and Perry [27] to introduction of *situation semantics*, in which situations are segments of possible worlds.

[23] Both Montague [258] and Tichý [419] employed terms for intensions and also extensions in their systems, while Tichý's explicit work with ws and ts has an advantage over Montague's $^\wedge$ and $^\vee$ (see Jespersen [193], where the reference to Montague is only implicit, however). THL shares Tichý's idea, but avoids its superfluous and irritating use.

equivalent to the next condition); $F/(\pi\tau)$; $X/\tau; TD/\tau_{\rho\omega}$ (an office equivalent to the next office); $D/\tau_{\omega\rho}; H/(o\tau)$:

$$\begin{array}{lcl}(\text{``}[\lambda w\lambda t[TF_{wt}\,X]]\text{''})^{\bullet} & = & \text{``}[F\,X]_{tw}\text{''}\\ (\text{``}[\lambda w\lambda t[TF_{wt}\,TD_{wt}]]\text{''})^{\bullet} & = & \text{``}[F\,D_{tw}]_{tw}\text{''}\\ (\text{``}[\lambda w\lambda t[H\,TD_{wt}]]\text{''})^{\bullet} & = & \text{``}[H\,D_{tw}]\text{''}\end{array}$$

But the translational rule $(\text{``}\cdot\text{''})^{\bullet}$ does not apply universally, some TIL analyses do not have a straightforward translation to THL. My first example below shows that it is because THL reaches extensions of certain intensions in a different way to TIL. The second example, which analyses the sentence

"Xenia was white.",

shows one of many differences between THL and TIL as regards analysis of verb tenses (I will return to the issue in the next point.) Let ${}^{(0)}\mathbf{TWa}, {}^{(0)}\,\mathbf{TH_2O}/(o\iota)_{\rho\omega}$ (a property as analysed by TIL); ${}^{(0)}\mathbf{Was}/(\pi\pi); t'/\rho$; ${}^{(0)}\mathbf{Pret}$ (preteritum) $/((o(o(o\rho))(o\rho))\rho)$; ${}^{(0)}\mathbf{Once}/(((o(o\rho))\pi)\omega)$.

Example 38 (Comparison of THL and TIL analyses II)

THL *analysis*	TIL *analysis*
$[[\lambda x[\mathbf{Wa}\,x]_{tw}] =^{(o\iota)} [\lambda x[\mathbf{H_2O}\,x]_{tw}]]$	$[\lambda w\lambda t[\mathbf{TWa}_{wt} =^{(o\iota)} \mathbf{TH_2O}_{wt}]]$
$[\mathbf{Was}\,[\mathbf{Wh}\,\mathbf{X}]]_{tw}$	$[\lambda w\lambda t[\mathbf{Pret}_t[\mathbf{Once}_w$ $[\lambda w\lambda t[\mathbf{TWh}_{wt}\,\mathbf{X}]][\lambda t'[t' =^{\rho} t]]]]]$

(iii) THL analyses of *verb tenses* differ from those of TIL. They are obviously more concise and they also better fit the structure of sentences.

To describe the difference, let me begin with the aforementioned THL analysis

$$[\mathbf{Was}\,[\mathbf{Wh}\,\mathbf{X}]]_{tw}$$

It contains the 'past tense operator', which is a medadic condition for propositions, while it maps the proposition THAT XENIA IS WHITE, which is v-constructed by $[\mathbf{Wh}\,\mathbf{X}]$, to the proposition that has T at the values of t and w iff there is a moment of time preceding the value of t in a given value of w such that THAT XENIA IS WHITE holds in it (being thus true); in all other cases it is false.[24]

The above construction is thus v-congruent with (where ${}^{(0)} < /(o\rho\rho)$; ${}^{(0)}\exists^{\rho}/(o(o^{\rho})))$:

$$[\exists^{\rho}\lambda t'[[t' < t] \wedge [\mathbf{Wh}\,\mathbf{X}]_{t'w}]]$$

[24] If it would be gappy when the input proposition was gappy, the next definiens of past tense operator should be appropriately modified.

This naturally fits our understanding of the past tense, so one defines (where p/π)

$$\models [\mathbf{Was}\, p]_{tw} \Leftrightarrow_o [\exists^\rho \lambda t'[[t' < t] \wedge p_{t'w}]]$$

For definitions of present perfect, present, future and future perfect operators it is enough to replace $<$ by $\leqslant, =^\rho, \leqslant$, and $>$, respectively.

In the case of sentences involving an expression for *reference interval*, e.g.

"On 1/1/1977, Xenia was white.",

their THL analyses can utilise the dyadic condition satisfiable by a proposition and reference intervals (where ${}^{(0)}\mathbf{Was2}/(\pi\pi(o\rho))$; ${}^{(0)}1/1/1977/(o\rho)$):

$$[\mathbf{Was2}\, [\mathbf{Wh}\, \mathbf{X}]\, 1/1/1977]_{tw}$$

which has the definition (where $i/(o\rho)$):

$$\models [\mathbf{Was2}\, p\, i]_{tw} \Leftrightarrow_o [\exists^\rho \lambda t'[[t' < t] \wedge p_{t'w} \wedge [i\, t']]]$$

Frequency adverbs such as e.g. "once", "twice", "intermittently" can be understood either as names of the third argument for an appropriate triadic condition, or as modifiers of the input proposition such as THAT XENIA WAS WHITE ON 1/1/1977.

On the other hand, Tichý (esp. [423], see [419]) tried to analyse the three mentioned types of sentences using only one template. To achieve it, he analysed sentences that do not involve phrases for reference intervals or frequency adverbs as equivalent to sentences involving their 'dull names', such as "once" (cf. e.g. the TIL analysis in point iii., comparison II).

Tichý's [423] analyses and definitions are rather complicated because he did not use conditions of THL, but TIL intensions. Cf. e.g. his definition of the 'past tense operator' (where $t''/\rho;{}^{(0)}=^o/(ooo);{}^{(0)}\imath^o/(o(oo)); j/(o(o\rho))$ – the value of j is the result of application of a frequency adverb on the base proposition):

$$\models [\mathbf{Pret}_t\, j\, i] \Leftrightarrow_o [\imath^o \lambda o[[\exists^\rho \lambda t'[[i\, t'] \wedge [t' < t]] \wedge [o =^o [j\, \lambda t''[[i\, t''] \wedge [t'' < t]]]]]]]$$

(iv) Unlike THL, TIL shares an important lack of expressive capability with PWS: it does not sufficiently take into consideration that an expression, especially a sentence, is sometimes used for reference to an extension, but sometimes to an intension, hence, both *extensional* and *intensional readings* (in the first sense used in this chapter) are to be provided.

Descriptions also make a good example. Arguably, in most contexts of everyday usage, a description is deployed for reference to an individual (or an object of another type). When Xenia utters "Xenia's dog is happy.", she normally intends to elicit attention towards the fact that her dog (say) Fido satisfies the condition in question. On the other hand, contexts in which the reference is rather virtual than factual also exist: when Xenia utters "The poorest man on Earth must be sad." (to evoke Frege's [142] well-known example about an African chief), she hardly makes a reference to a particular individual, the description instead stands for an individual office.

Now consider sentences. My argument will use an analysis of simple (cf. 4.5.1) and belief sentences (5) both in TIL and THL. The usual purpose of an ordinary utterance of a sentence such as

S: "The President of the USA is white."

is to refer to a(n unspecified) truth value, i.e. an extension. If, on the other hand, *S* occurs nested in a belief context, e.g.,

BS: "Xenia believes that the President of the USA is white."

its usual purpose is to report the content of Xenia's belief, i.e. a 'proposition'.

PWS explicated 'propositions' as propositions. If accepting PWS, a choice must then be made. Either one decides to i. accept contextualism according to which *S* denotes an intension if it occurs in an (hyper)intensional context, while *S* refers to a truth value if it occurs in an extensional context, or one decides to ii. employ propositions systematically, regardless of the context in which *S* occurs. A distinctive variant of such a contextualism is Carnap's [57] Method of Extension and Intension; followed by Montague [258]. Option ii. has also often been adopted within PWS.

Tichý implemented the 'anti-contextualistic' ii.-version of PWS in his unpublished book [418] and a series of his published papers [419]. In [422], he upgraded it to the hyperintensional level: *S* stands for a certain construction of a proposition in both direct and indirect context. Using TIL, Tichý analysed *S* and *BS* as expressing

$$S^{\mathsf{TIL}} \quad [\lambda w \lambda t[\mathbf{TWh}_{wt}\, \mathbf{TPrU}_{wt}]]$$
$$BS^{\mathsf{TIL}} \quad [\lambda w \lambda t[\mathbf{TBel}^1\, \mathbf{X}\, {}^0[\lambda w \lambda t[\mathbf{TWh}_{wt}\, \mathbf{TPrU}_{wt}]]]],$$

respectively, where the second relatum of the belief attitude is delivered by its immediate construction, its 0-execution; see 5.3 for discussion of just this point.

On the other hand, THL preserves the aforementioned intuition that, depending on context, a sentence is either used for reference to an extension, or to an intension (or even a hyperintension). Belief sentences are analysed as reporting a relation(-in-intension) between agents and constructions of a truth value. A sentence reporting an object of someone's belief attitude expresses such a construction both in belief and ordinary contexts. What changes, however, is the way in which the construction is locked into the construction of which it is a subconstruction (whether it is combined with t and w, or placed in the scope of 0-execution, etc.).

In THL, S and BS are analysed (where ${}^{(0)}\mathbf{Bel}^1/(\pi\iota *_1)$ – a dyadic condition satisfiable by individuals and 1st-order constructions) as

$$S^{\mathsf{THL}} \qquad [\mathbf{Wh}\,\mathbf{PrU}_{tw}]_{tw}$$
$$BS^{\mathsf{THL}} \qquad [\mathbf{Bel}^1\,\mathbf{X}\,{}^0([\mathbf{Wh}\,\mathbf{PrU}_{tw}]_{tw})]_{tw}$$

respectively. (Round brackets are used as an auxiliary to indicate the scope of 0-execution.) Note that it is the employment of 0-execution that enables us to get rid of the PWS idea that S stands for the proposition in intensional (and then also extensional) context.

(v) At first sight, TIL is more systematic than THL, since a sentence such as S is uniformly treated as expressing a propositional construction. As discussed above, this advantage comes with the price of sweeping aside the intuition that S is normally used for reference to a(n unspecified) truth value. But the impression of the uniformity of TIL analyses is also questionable.

First, note that Tichý's TIL brings a complication in analysis of mathematical sentences. An ordinary mathematical sentence such as

$$M \qquad \text{“}1+2=3\text{”}$$

is analysed by Tichý [422] in a rather ponderous manner, while THL offers a neat and straightforward analysis, cf.

$$M^{\mathsf{TIL}} \qquad [\lambda w \lambda t[[1+2] =^{\rho} 3]]$$
$$M^{\mathsf{THL}} \qquad [[1+2] =^{\rho} 3]$$

Now imagine an attempt to define (say) the notion of truth that is applicable to constructions. In THL, one would naturally consider o-constructions as the bearers of the property/condition BE A TRUE CONSTRUCTION. In TIL, however, the primary bearers of such a property/condition are propositional constructions. The notion of truth applicable to o-constructions is then defined either as somehow parasitic, or rather as independent on the former notion.[25] TIL thus enforces an unnecessary dualism on the side of the analyst, i.e. contextualism has been shifted elsewhere.

v. Observe also that, contrary to appearance, THL incorporates the intuition according to which a sentence excludes a range of possibilities, as is usually attributed to PWS. In PWS or TIL, the intuition is implemented on the level of denotata: S excludes more possibilities than S' if the proposition

[25] See Raclavský [323] for an analysis of truth in TIL. TIL analysis of entailment, offered in Raclavský, Kuchyňka and Pezlar [339], suffers from a similar problem.

denoted by S contains less world-time couples than the proposition denoted by S'.

In THL, the intuition is not handled on the level of intensions, but hyperintensions. S excludes more possibilities than S' if the construction expressed by S excludes more (equivalence sets of) constructions as incompatible. This is a deduction-related procedure: the construction $[[1+2] =^{\rho} 3]$, for example, excludes $[\neg[[1+2] =^{\rho} 3]]$ and also e.g. $[\neg[[2+1] =^{\rho} 3]]$, depending on the derivation rules governing the subconstructions of the compared constructions.

Chapter 5

Belief and substitution

5.1 Overview of the chapter

Recall from chapter 1 that *belief sentences* typically have the "X Vs O." scheme, where the belief operator "V" is a transitive verb that connects an expression for an agent X with an expression for the *object* O of the belief attitude reported by the sentence. Thus, belief sentences belong to a large family of sentences reporting *intentional attitudes*.[1]

Both belief attitudes and the other types of attitudes are governed by the very similar group of derivation rules, and so my brief examination of the latter type of attitudes will put my investigation of belief attitudes into a wider context.

The semantics of sentences reporting attitudes should go hand in hand with intuitively acceptable inferences involving "O" and reasonable ontological assumptions concerning O. Many of Vs contained in such sentences, especially "think about", "contemplate", etc., are highly ambivalent with regards to their logical type since the Os are of various types. I split the sentences/attitudes into three main groups, depending on whether O's type is

a. a common extension (i.e. a non-construction and non-intension; 5.2.1),

b. an intension (5.2.2), or

c. a construction (5.3).

Attitudes towards intensions form a large group of attitudes (the corresponding verbs are widely known as *intensional transitives*). Among them, I recognise *attitudes towards propositions*; I do not call them "propositional attitudes" so as to avoid confusion with the analysis of belief attitudes.

---

[1] Forbes [132] recently offered another apt term for intentional attitudes, "*objectual attitudes*". See his book [132] for a comprehensive investigation of the topic. The first part of this chapter in particular utilises results from Raclavský [328] and Raclavský, Kuchyňka and Pezlar [339].

Belief attitudes are not explicated as attitudes towards possible world propositions, but towards constructions of truth values, i.e. o-constructions (5.3). Recalling Levesque's [234] famous implicit–explicit belief distinction, belief attitudes are analysed in THL as *explicit beliefs* (the apparent restrictiveness of this approach is thoroughly discussed in the next chapter, 6). The motivation for such an explication was stated in the introductory chapter 1: the adoption of hyperintensional, fine-grained, algorithmic meanings capable of avoiding the Paradox of Hyperintensional Contexts as well as the Paradox of Logical Omniscience (6.2).

In this chapter, I focus on *de dicto* and *de re* readings of belief sentences (5.4.2) as occurring in valid/invalid arguments. I examine which substitutions in them are possible (5.4.3). I show that an expression may syntactically occur in the scope of a hyperintensional operator, but that it does not automatically mean that it occurs in a genuine hyperintensional context (5.4.4).

5.2 Attitudes and substitution

5.2.1 Attitudes towards common extensions

First, I provide two analyses of attitudes towards extensions (where $w/\omega; t/\rho$; ${}^{(0)}\mathbf{X}$ (Xenia), ${}^{(0)}\mathbf{Y}$ (Yannis) $/\iota$; ${}^{(0)}\mathbf{Pr}$ (be a prime number)$/(o\rho)$; ${}^{(0)}\mathbf{Cont}^{\iota}$ (contemplate) $/(\pi\iota\iota)$; ${}^{(0)}\mathbf{Cont}^{(o\rho)}/(\pi\iota(o\rho))$):

Example 39 (Attitudes towards common extensions)

"Xenia contemplates Yannis."	$[\mathbf{Cont}^{\iota}\,\mathbf{X}\,\mathbf{Y}]_{tw}$
"Xenia contemplates primes."	$[\mathbf{Cont}^{(o\rho)}\,\mathbf{X}\,\mathbf{Pr}]_{tw}$

Analyses of sentences reporting attitudes must meet intuitively valid inferences involving such sentences, the most important ones are (the intuitive versions of) (SI) and (EG). We will repeatedly see that correct applications of various derivation rules governing attitude reports embody the following, intuitively obvious methodological rule:

> *'Golden Rule' of Attitude Logic (GRAL).* The correctness of the substitution in sentences that report attitudes stems from the fact that one does not change the object of an attitude that is ascribed to an agent; one may only change in which way the object is referred.

An object O of an attitude can be determined by means of O's name, or in an indirect way: by means of some O's description "D". To simplify the wording of the following consideration, let O and D, D' (as well as W, T, A, R, R^2 below) be considered directly as constructions expressed by "O", "D", "D", where

$O/\tau; D, D'/\tau_{\omega\rho}$. A substitution of D for O in a particular argument may require D be 'applied to' the constructions T and W (where $T/\rho; W/\omega$).

The next schemes are particular derivation schemes licensed by (SI) or (EG) (3.4.3), which are known to fail in their ordinary formulations (I will return to this issue in the next section). Let R be a construction of some type-theoretically appropriate attitude as dyadic condition, i.e. $R/(\pi\iota\tau)$, and A be a construction of an agent, i.e. A/ι; further, let x/τ. Instead of O, there might occur a certain D' that would be 'applied to' T and W, i.e. D'_{TW}.

Example 40 (Valid schemes licensed by (SI) and (EG) I)

$$\frac{(R(A, O))_{TW} \qquad O = D_{TW}}{(R(A, D_{TW}))_{TW}}\ \text{(SI)} \qquad\qquad \frac{(R(A, O))_{TW}}{\exists x(R(A, x))_{TW}}\ \text{(EG)}$$

Here are straightforward instances of the derivation schemes:

Example 41 (Valid arguments licensed by (SI) and (EG) I)

$$\frac{\text{“Xenia contemplates Yannis.”} \qquad \text{“Yannis = Zoë's best friend.”}}{\text{“Xenia contemplates Zoë's best friend.”}}\ \text{(SI)}$$

$$\frac{\text{“Xenia contemplates Yannis.”}}{\text{“There is an individual such that Xenia contemplates him.”}}\ \text{(EG)}$$

Such arguments can be easily formalised following exemplary analyses in the previous 4 and current chapters. Moreover, it is easy to check their validity either syntactically, using (SI) and (EG) rules offered in 3.1, or semantically, by computing truth-preservation through the (formalisation of the) argument. (Similarly below.)

5.2.2 Attitudes towards intensions

i. 'Notional attitudes'

As is well known, Quine [315] noticed that the above 'relational reading' of sentences with transitive verbs – which treats them as reporting attitudes towards ordinary extensions, such as individuals – is untenable if the object of an attitude is a 'notion'. Numerous verbs encoding such attitudes have been listed as examples: “seek”, “worship”, “admire”, “afraid of”, “imagine”, etc.

I follow Tichý [418] in using Quine's term "notional reading" to call such attitudes "*notional attitudes*", though the topic is often discussed under Montague's [260] term "*intensional transitives*" (*ITVs*). I stick to Tichý's terminology here, since the term "ITV" applies to verbs, not attitudes, while it may also cover belief verbs, which I do not want.

When analysing sentences reporting notional attitudes, Quine observed that both classical principles (SI) and (EG) fail if they are applied to them. Montague (e.g. [260, 263]) reacted to the problem by offering his famous analysis that deploys PWS.[2] According to Montague, as well as Tichý [410], who adapted Montague's analysis, the object of a notional attitude is not the object (if any) referred to by the description "D", but an office (Montague: "individual concept") denoted by "D", i.e. an intension.

A logical analysis of a sentence reporting a notional attitude contains D that is not 'applied to' T and W; D thus serves to deliver the office as such, not its value in a particular T and W. Precisely this solution appears to be Tichý's original contribution ([410], see also [418, 422]); Montague worked with the intensional operator "$^\wedge$" to the similar effect. Of course, the idea is generalised even for cases of attitudes towards properties, relations(-in-intension), etc.[3]

Here are examples (where ${}^{(0)}\mathbf{L}$ (León) $/\iota$; ${}^{(0)}\mathbf{Seek}/(\pi\iota\iota_{\omega\rho})$; ${}^{(0)}\mathbf{Fy}/\iota_{\omega\rho}$ (the Fountain of Youth) – which is a simplification; ${}^{(0)}\mathbf{Uni}$ (be a unicorn) $/(\pi\iota)$; ${}^{(0)}\mathbf{Cont}^{(\pi\iota)}/(\pi\iota(\pi\iota)))$:

Example 42 (Notional attitudes)

"Ponce de León seeks the Fountain of Youth."	$[\mathbf{Seek}\,\mathbf{L}\,\mathbf{Fy}]_{tw}$
"Xenia contemplates unicorns."	$[\mathbf{Cont}^{(\pi\iota)}\,\mathbf{X}\,\mathbf{Uni}]_{tw}$

Remarks. Richard's [347] well-known objection against Montaguean analysis of notional attitudes, which is based on the use of valid arguments such as "Xenie seeks a white unicorn. Therefore, Xenia seeks a unicorn.", is avoided in the present approach (see details in Raclavský, Kuchyňka and Pezlar [339]), because the sentences in such arguments are analysed as quantified sentences that report relations(-in-intension) of agents to offices that satisfy a certain property (e.g., that a holder of each such sought office must be a unicorn). That is, one is not forced, as in Montague's approach, to analyse the just mentioned argument as reporting two distinct attitudes to two different properties, which would render the argument as invalid.

[2] Quine's [315] own analysis of notional attitudes is generally deemed unsatisfactory, see e.g. Forbes [132]. It should be added that Montaguean analysis resembles Church's [67] solution, for whom the object of an attitude is an individual concept of Church's logical type ι_1 (while individuals are of type ι_0). Montague should be credited for modelling these individual concepts as possible world intensions.

[3] Tichý's analyses of notional attitudes have been re-examined by his followers repeatedly, e.g. Raclavský [328], Duží, Jespersen and Materna [110].

ii. Attitudes of wishing

Analogous behaviour is also shown by other types of attitudes towards intensions, e.g. the attitudes of wishing, which I expose here even for a subsequent comparison with belief attitudes.

According to THL, sentences such as

"Xenia wishes to become the President of the USA.",

report *attitudes towards conditions.* I thus follow the proposals by Montague [260], Oddie and Tichý [278] who analysed them as reporting attitudes towards properties. There is a plenitude of verbs playing a part in such wishing sentences: "desire", "intend", "require", "prefer", "refute", etc.

On the other hand, sentences reporting that an agent wishes that somebody else satisfies a certain monadic condition are analysed rather as sentences reporting an *attitude towards propositions*, i.e. towards medadic conditions. On that reading, the agent wishes a certain state-of-affairs (fact).

Here are examples (where ${}^{(0)}\mathbf{PrU}$ (the President of the USA) $/\iota_{\omega\rho}$ – which is a simplification; ${}^{(0)}\mathbf{WishTo}/(\pi\iota(\pi\iota))$;${}^{(0)}$ **Wish** (wish [that])$/(\pi\iota\pi)$; ${}^{(0)}\mathbf{Be}^{\iota}/(\pi\iota\iota_{\omega\rho})$ – see 4.4.3):

Example 43 (Wishing attitudes towards conditions and propositions)

"Xenia wishes to become the President of the USA."	$[\mathbf{WishTo}\,\mathbf{X}\,[\lambda x[\mathbf{Be}^{\iota}\,x\,\mathbf{PrU}]]]_{tw}$
"Xenia wishes that Yannis was the President of the USA."	$[\mathbf{Wish}\,\mathbf{X}\,[\mathbf{Be}^{\iota}\,\mathbf{Y}\,\mathbf{PrU}]]_{tw}$

Remark. There are famous examples showing *de re–de dicto (scope) ambiguity* of sentences such as "Oedipus wishes to marry his mother.", which is a straightforward analogue of the well-known modal *de dicto–de re* ambiguity. The ambiguity is easily removable in THL by properly employing variables for possible worlds and moments of time that are free or bound in the whole construction.[4]

5.2.3 Substitution and attitudes towards intensions

As a result of his indisposition to distinguish descriptions from proper names, Quine [316] did not recognize the obvious reason why classical formulations of (SI) and (EG) fail. The failure of (SI) happens if a. one substitutes on the basis of congruence $D_{TW} = D'_{TW}$ (on the language level: the co-reference of

[4] By means of TIL, the particular example was analysed e.g. by Kolář and Svoboda [215].

two descriptions) but b. the object of an attitude attributed to an agent is thus changed – which is a violation of GRAL. In the case of (EG), one derives the existence of something that could be a mere chimera, so to speak, because the description may refer to anything at all (again, one would change the object of an agent's attitude).

Here is a list of invalid derivation schemes (and particular arguments) which are only seemingly licensed by the rules (SI) and (EG), while they are actually not, which is why I mark them with "~~(SI)~~" and "~~(EG)~~".

Example 44 (Invalid schemes unlicensed by (SI) and (EG))

$$\frac{\begin{array}{c}(R(A,D))_{TW}\\ D_{TW} = D'_{TW}\end{array}}{(R(A,D'))_{TW}}\ \cancel{(SI)} \qquad\qquad \frac{(R(A,D))_{TW}}{\exists x(R(A,x))_{TW}}\ \cancel{(EG)}$$

Example 45 (Invalid arguments unlicensed by (SI) and (EG))

$$\frac{\begin{array}{c}\text{“Ponce de León seeks the Fountain of Youth.”}\\ \text{“The Fountain of Youth = Boinca's fountain of youth.”}\end{array}}{\text{“Ponce de León seeks Boinca's fountain of youth.”}}\ \cancel{(SI)}$$

$$\frac{\text{“Ponce de León seeks the Fountain of Youth.”}}{\text{“There is an individual such that Ponce de León seeks it.”}}\ \cancel{(EG)}$$

Contrary to popular belief, I maintain that the principles (SI) and (EG) work properly even when applied to sentences reporting notional attitudes. However, distinct elements must be involved. When applying (SI), a construction that is expressed by a description referring to an office, i.e. $D^2/\tau_{\omega\rho\omega\rho}$, plays a part in the argument. In the case of the application of (EG), one needs a variable for offices, $d/\tau_{\omega\rho}$. The resulting valid derivation schemes are nothing but appropriate type-theoretic modifications of the valid schemes of type I (5.2.1).

Example 46 (Valid schemes licensed by (SI) and (EG) II)

$$\frac{\begin{array}{c}(R(A,D))_{TW}\\ D = D^2_{TW}\end{array}}{(R(A,D^2_{TW})_{TW}}\ (SI) \qquad\qquad \frac{(R(A,D))_{TW}}{\exists d(R(A,d))_{TW}}\ (EG)$$

To illustrate the valid scheme licensed by (SI) of type II, consider that the object of someone's attitude is an office, say THE FOUNTAIN OF YOUTH, which is referred to not by "the Fountain of Youth", but by a 'higher-order' description such as "the favourite office of [Barack] Obama", which denotes an office of type $\iota_{\omega\rho\omega\rho}$. The example illustrating (EG) of type II also sounds somewhat artificial, but again, it may occur in philosophical discourse.

Example 47 (Valid arguments licensed by (SI) II)

"Xenia contemplates the Fountain of Youth."
"The Fountain of Youth is the favourite office of Obama."
——————————————————————— (SI)
"Xenia contemplates the favourite office of Obama."

"Ponce de León contemplates the Fountain of Youth."
——————————————————————— (EG)
"There is an individual office of the Fountain of Youth such that Ponce de León contemplates it."

5.3 Attitudes towards constructions

Sentences are typically analysed as reporting attitudes towards constructions if the object of an attitude is a complex 'procedure', not the outcome of the execution of such a 'procedure'. For satisfactory analysis (which also satisfies GRAL), each computing 'procedure', i.e. a construction C, is introduced in the whole analysis by means of its 0-execution, for ^0C v-constructs the 'procedure'-construction C as such, not its result (where $^{(0)}64, ^{(0)}3, ^{(0)}0/\rho; ^{(0)}\surd/(\rho\rho);\ ^{(0)}\div/(\rho\rho\rho); ^{(0)}\,\mathbf{Calc}/(\pi\iota*_1))$:[5]

Example 48 (Attitudes towards constructions)

"Xenia calculates the square root of 64."	$[\mathbf{Calc}\,\mathbf{X}\,^0[\surd 64]]_{tw}$
"Xenia calculates 3 ÷ 0."	$[\mathbf{Calc}\,\mathbf{X}\,^0[3 \div 0]]_{tw}$

Remark. Obviously, such analyses cannot be achieved within STT or any other 1st-order framework, since one deploys here 2nd-order constructions of certain 1st-order constructions, for which RTT is needed.

The most prominent attitudes towards constructions are *belief attitudes.*

Definition 80 (Belief attitudes)

Belief attitudes are attitudes explicated as dyadic conditions satisfiable by agents and o-constructions that are meanings of the embedded sentences of belief sentences, i.e. as objects of type $(\pi\iota*_k)$.

[5] Such examples were analysed by means of TIL by Materna [248] on the basis of Tichý's [422] analysis of "3 ÷ 0 is not a proper construction." (2.6.3).

Belief attitudes intuitively differ from attitudes towards propositions. To illustrate, when Xenia wishes that Yannis was the President of the USA, it is not important whether one refers to the desired state as a state-of-affairs in which

(a) Yannis is the President of the USA, or

(b) Yannis is the President of the USA and Fermat's Last Theorem (FLT) holds.

In the case of belief attitudes, on the other hand, one does not have such a choice of co-denotative expressions, for there is a difference between whether one ascribes to Xenia that she believes (a), or (b).

Here is the analysis of the second indicated belief sentence on its *de dicto* reading (where $x, y, z, n/\rho; {}^{(0)}< /(o\rho\rho); {}^{(0)}\wedge, {}^{(0)}\rightarrow /(ooo); {}^{(0)}\forall^{\rho}/(o(o\rho)); {}^{(0)}=^{\iota} / (o\iota\iota); {}^{(0)}=^{\rho} /(o\rho\rho); {}^{(0)}\mathbf{Bel}^1$ (believe) $/(\pi\iota *_1)$):

Example 49 (Semantics of *de dicto* belief sentence)

$$\text{“Xenia believes that Yannis is the President of the USA and } \forall x \forall y \forall z \forall n((x^n + y^n = z^n) \rightarrow (n < 3)).\text{”} \mid$$

$$[\mathbf{Bel}^1\,\mathbf{X}\,{}^0([[\mathbf{Y} =^{\iota} \mathbf{PrU}_{tw}] \wedge [\forall^{\rho}\lambda x \forall^{\rho}\lambda y \forall^{\rho}\lambda z \forall^{\rho}\lambda n[[[x^n + y^n] =^{\rho} z^n] \rightarrow [n < 3]]]])]_{tw}$$

Remarks. In the record of the construction analysing the expression of FLT I suppress several pairs of brackets. I add round brackets to indicate the scope of the displayed 0-execution; similarly below. According to my analysis, the agent believes a 1st-order construction; I will focus on the relationship of beliefs to orders in 5.5.2. The contrast between *de dicto* and *de re* beliefs will be studied in 5.4.1 and 5.4.2.

Comments.

i. Tichý [422] suggested such a style of analysis when addressing the Principle of Subject Matter (Aboutness, 4.2.1). According to him, however, belief attitudes are attitudes towards propositional constructions, not towards o-constructions.

> [the PWS] theory [of aboutness] ... portrays the [maths] teacher as never mentioning the item which is at the heart of the matter [i.e. mathematics] and which he is anxious to bring to the pupil's attentions. It does not impute the teacher the reference to the *truth-value* determined by the proposition constructed by $\lambda w \lambda t. = [+\,\mathbf{1}\,\mathbf{1}]\mathbf{2}$, but to the *proposition itself*. But the trivial proposition (the unique proposition which is true in all worlds at all times) is no more the subject matter of ['One plus one makes two'] than it is the truth-value T. The real subject

matter that the sentence treats of–namely the construction $\lambda w \lambda t. = [+ \mathbf{1}\,\mathbf{1}]\,\mathbf{2}$–goes on my own modification of Frege's theory [of reference], unnoticed

Tichý [422], pp. 223–224

As discussed in 4.6, Tichý's proposal preserves one undesirable feature of the PWS analysis of belief attitudes: it is incapable of discriminating between (say)

(i) "Fido is a dog" and

(ii) "that Fido is a dog",

i.e. between extensional and intensional uses of the sentence "Fido is a dog.". In THL, the object of belief is an o-construction of form C_{tw}, since (i) is normally used to speak about a(n unspecified) truth value, not about a set of time-world couples constructed by C, which is the purpose of uttering (ii).

ii. One of the crucial requirements is fulfilled in the THL analysis: one avoids the Paradox of Logical Omniscience (see 6.2 for discussion) since ascriptions of belief to other objects than the agent was originally believing (recall GRAL) is prevented. Substitution for indirect subconstructions (3.4.2) is not permitted. Though

$$^0[[\mathbf{Y} =^{\iota} \mathbf{PrU}_{tw}] \wedge [\forall^{\rho}\lambda x \forall^{\rho}\lambda y \forall^{\rho}\lambda z \forall^{\rho}\lambda n[[x^n + y^n =^{\rho} z^n] \rightarrow [n < 3]]]]$$

is a direct subconstruction of the whole construction of the above example, no subconstruction of this construction is a direct subconstruction of the whole construction, and so it is not possible to substitute another construction of the number 3 for (say) the variable in place of $^{(0)}3$. (I will return to the substitutivity issues in 5.4.3 below.) The agent's belief is explained as the *explicit attitude* towards the construction introduced by its 0-execution.

iii. Also observe that the THL style of analysis incorporates Frege's [143] distinction of *direct/indirect sense* ("gerade/ungerade Sinn"): the meaning of a sentence, i.e. its direct sense ("gerade Sinn"), is a certain construction C, while a truth value (if any) v-constructed by C is its denotatum ("gerade Bedeutung"). If the sentence is introduced in an indirect context, i.e. it is embedded in a belief sentence, C becomes its indirect denotatum ("ungerade Bedeutung"); the 0-execution of C, viz. ^0C, becomes its indirect sense ("ungerade Sinn").

Remark. One might perhaps say that such an analysis is contextualistic in the positive sense of the word. The indirect context indicated by "that" leads us in the context of the analysis of the whole sentence to a 'lift' of the meaning of the embedded sentence from C to ^0C. To avoid possible misunderstanding, the meaning of the embedded sentence is neither C nor, sometimes, in an indirect context, ^0C; the meaning of the embedded sentence is still just and only C – but C can be locked in another construction in various ways, either simply as C, or within the scope of its 0-execution, i.e. ${}^0C$, or perhaps in another way.

iv. THL is also capable of treating *illogical beliefs* (Raclavský [328]). It is not limited to the case of believing contradictory constructions such as $[\neg[1 =^\rho 1]]$, for e.g. Xenia may believe that $[[\lambda x[xx]][\lambda x[xx]]]$ yields (say) the Sun since she shares *horror vacui* concerning the partiality gap with Frege [139]. For $[[\lambda x[xx]][\lambda x[xx]]]$ is v-improper, she fancies the construction rather v-constructs a selected object from the universe, as Frege propagated (where ${}^{(0)}\mathbf{Sun}/\iota; x/(\rho\rho)$):[6]

Example 50 (Illogical belief)

"Xenia believes that $[[\lambda x[xx]][\lambda x[xx]]]$ = the Sun."	$[\mathbf{Bel}^2\ \mathbf{X}\ {}^0[{}^{20}[[\lambda x[xx]][\lambda x[xx]]] =^\iota \mathbf{Sun}]]_{tw}$

Remark. In early intuitionistic/constructive type theory, the approach to belief attitudes was rather different: knowledge and belief were considered to be dispositions, while knowledge was acquired by an act of judgement, see e.g. Martin-Löf [246], Ranta [342]. A comprehensive theory of belief attitudes for constructive type theory was recently developed by Wieckowski [448].

5.4 Belief *de dicto/de re* and substitution

5.4.1 Belief *de dicto/de re* and PWS

In chapter 1, I introduced the well-known *de dicto–de re* distinction that captures two different readings of belief sentences, which are subsequently called *beliefs de dicto* and *beliefs de re*. It enables the blocking/allowing of certain inferences that are based on the equivocation of the *de dicto* and *de re* readings of belief sentences.

To illustrate, the next two arguments are valid if the belief sentences involved in them are read as beliefs *de re*, which is their most natural reading here.

---

6 Note that when a $\mathcal{L}_{\mathsf{TTT}}$-term "$C$" occurs in an English sentence S, one usually analyses the sentence by means of a construction D that contains the 0-execution of the construction C that is displayed by "C"; "C" 'mentions' C and so C is not executed (cf. Raclavský [326]). In the above example, however, "C" is used to refer to the value (if any) of C, so C must occur in D as executed, which is why I utilise the 2-execution of ${}^0C$, i.e. ${}^{20}C$; ${}^{20}C$ is v-congruent with C.

Example 51 (Valid arguments involving belief *de re* that are licensed by (SI) and (EG))

$$\frac{\text{“Xenia believes of the President of the USA that he is blue-eyed.”} \quad \text{“The President of the USA is the first African-American president.”}}{\text{“Xenia believes of the first African-American president of the USA that he is blue-eyed.”}}\ (\mathrm{SI})$$

$$\frac{\text{“Xenia believes of the President of the USA that he is blue-eyed.”}}{\text{“There exists somebody such that Xenia believes that he is blue-eyed.”}}\ (\mathrm{EG})$$

In contrast, the following argument appears to be intuitively invalid. From this fact one should conclude that both its belief sentences are to be read in the *de dicto* way (not the *de re* way, on which the argument is valid).

Example 52 (Invalid argument blocked within PWS and THL by employing *de dicto* reading)

$$\frac{\text{“Xenia believes that the President of the USA is blue-eyed.”} \quad \text{“The President of the USA = the first African-American president.”}}{\text{“Xenia believes that the first African-American president is blue-eyed.”}}$$

Similarly to PWS, THL utilises the *de dicto–de re* distinction. But since its implementation in THL requires it to surpass a certain technical obstacle (5.4.2), I first describe its implementation in a certain THL-version of PWS. Let us temporarily admit that the objects of beliefs are propositions; then, the two readings of the major premise of the above argument are formalised as follows (where $w'/\omega; t'/\rho$ ${}^{(0)}\mathbf{Bl}$ (be blue-eyed) $/(\pi\iota);{}^{(0)}\,\mathbf{Bel}^{\pi}/(\pi\iota\pi)$).[7]

Example 53 (Belief *de dicto* and *de re* according to PWS)

belief *de dicto*	$[\mathbf{Bel}^{\pi}\ \mathbf{X}\ [\lambda t'\lambda w'[\mathbf{Bl}\,\mathbf{PrU}_{t'w'}]_{t'w'}]]_{tw}$
belief *de re*	$[\mathbf{Bel}^{\pi}\ \mathbf{X}\ [\mathbf{Bl}\,\mathbf{PrU}_{tw}]]_{tw}$

[7] Such analyses, but employing relations-in-intension instead of dyadic conditions, can be found in Tichý [410, 418]. Comparable analyses can be found e.g. in Montague [258], Hintikka [178] and works by their followers.

In the case of belief *de re*, the meaning of the description is combined with the free variables t and w because the occupant (if any) of the office denoted by the description plays a role in the verification of the whole belief sentence. The speaker uses such a sentence to express that Xenia believes of a certain individual, who is currently the President of the USA, to be blue-eyed. Thus, the sentence has the *existential presupposition* (see Strawson [393], Soames [381], Raclavský [332]) that the President of the USA exists (if the presupposition is not satisfied, the sentence lacks a truth value). It does not work this way in the case of belief *de dicto*: in order to verify the sentence, it is irrelevant who the President of the USA is (if any).

To explain the *de dicto–de re* distinction, Hintikka [179] drew an epistemological contrast between

a. the agent's perspective, when the reference of the substituted expression such as "the President of USA" is enclosed within the epistemic world of the agent, and

b. the speaker's perspective, when the reference of (in our case) "the President of USA" pertains to the epistemic world of the speaker.

There is no logical guarantee that the two perspectives overlap.

With regards to a., any substitution in the object of someone's belief would violate GRAL because the object of her attitude would change, and so one would unwarrantedly interfere in the area of the agent's epistemic competence. As regards b., on the other hand, substitution only changes the way the speaker refers to the object of the agent's attitude, which is warranted by the speaker's epistemic competence. Case b. thus does not differ from the case of the rather simple substitutions of (say) "the first African-American president of the USA" or "Obama" for "the President of the USA" into ordinary sentences such as "The President of the USA is blue-eyed.".

5.4.2 Belief *de dicto*/*de re* and THL

A certain obstacle in providing a correct THL analysis of belief sentences on their *de re* reading must be overcome. Recall that the THL analysis of *de dicto* reading of belief sentences, such as

"Xenia believes that the President of the USA is blue-eyed.",

'presents' construction C, which is the object of the belief attitude, by means of its 0-execution, i.e. ${}^0C$. Thus, any substitution in C is blocked.

This situation creates a problem if one attempts to transform such an analysis into an analysis of the corresponding *de re* reading, because, in it, the substitution in C must be allowed. For a correct analysis of the belief sentence in the *de re* reading, whose form is (where x/ι)

$$[\mathbf{Bel}^1\, \mathbf{X}\, {}^0([\mathbf{Bl}\, x]_{tw})]]_{tw}$$

one needs to replace x by the v-congruent construction $\mathbf{PrU}_{tw}$ using the *substitution function* SUB^k (3.4).

Recall that SUB^k assigns a construction C_4 to a ternary string ('triple') of constructions C_1, C_2, C_3, i.e. C_1 replaces (so to speak) the type-theoretically appropriate variable C_2 in C_3, while the result is C_4. In our case, C_1 would be the 0-execution of an individual (if any) who is v-constructed by the construction $\mathbf{PrU}_{tw}$. The individual, say Obama, is mapped to its 0-execution by the *trivialisation function* TRIV^1, i.e. ${}^{(0)}\mathbf{Triv}^{(*_1\iota)}/(*_1\iota)$ (cf. 2.5.1). Let ${}^{(0)}\mathbf{Sub}^1/(*_1 *_1 *_1 *_1)$:

$$[\mathbf{Sub}^1\, [\mathbf{Triv}^{(*_1\iota)}\, \mathbf{PrU}_{tw}]\, {}^0x\, {}^0([\mathbf{Bl}\, x]_{tw})],$$

where x corresponds to "he" in

"Xenia believes of the President of the USA that he is blue-eyed.".

Now we are equipped to illustrate the difference between the THL *de dicto* and *de re* readings of belief sentences.

Example 54 (Belief *de dicto* and *de re*)

belief *de dicto*	$[\mathbf{Bel}^1\, \mathbf{X}\, {}^0([\mathbf{Bl}\, \mathbf{PrU}_{tw}]_{tw})]_{tw}$
belief *de re*	$[\mathbf{Bel}^1\, \mathbf{X}\, [\mathbf{Sub}^1\, [\mathbf{Triv}^{(*_1\iota)}\, \mathbf{PrU}_{tw}]\, {}^0x\, {}^0([\mathbf{Bl}\, x]_{tw})]]_{tw}$

5.4.3 Substitution in *de dicto/de re* belief contexts

Now I am going to illustrate the difference between *de dicto* and *de re* by analysing an argument that is invalid on its *de dicto* reading, while it is valid on its *de re* reading, since only the *de re* reading of the belief sentences included in the argument allows substitution, which is needed for the establishment of its validity. The particular example I analyse is Example 52.

b.-*case*.

The *de re* reading of the argument (where ${}^{(0)}\mathbf{PrA}_{tw}$ (the first African-American president of the USA) $/\iota_{\omega\rho}$ – which is a simplification):

Example 55 (Substitution in belief *de re*)

$$\frac{\begin{array}{c}[\mathbf{Bel}^1\,\mathbf{X}\,[\mathbf{Sub}^1\,[\mathbf{Triv}^{(*_1\iota)}\,\mathbf{PrU}_{tw}]\,{}^0x\,{}^0([\mathbf{Bl}\,x]_{tw})]]_{tw}\\ [\mathbf{PrU}_{tw} =^{\iota} \mathbf{PrA}_{tw}]\end{array}}{[\mathbf{Bel}^1\,\mathbf{X}\,[\mathbf{Sub}^1\,[\mathbf{Triv}^{(*_1\iota)}\,\mathbf{PrA}_{tw}]\,{}^0x\,{}^0([\mathbf{Bl}\,x]_{tw})]]_{tw}}\ \text{(SI)}$$

is licensed by the rule (SI), since these three constructions C_1, C_2, C_3 are v-congruent, in the appropriate order, to the following constructions C_1', C_2, C_3', where C_1' and C_3' are logical forms of C_1 and C_3 (for simplicity only partial logical forms), respectively, and C_1', C_2, C_3' present an instance of (SI) (where ${}^{(0)}\mathbf{Sub}^2/(*_2\,*_2\,*_2*_2); {}^{(0)}\mathbf{Triv}^{(*_2\iota)}/(*_2\iota)$):[8]

$$\text{(SI)}\ \frac{\begin{array}{c}{}^2[\mathbf{Sub}^2\,{}^0(\mathbf{PrU}_{tw})\,{}^0y\,{}^0([\mathbf{Bel}^1\,\mathbf{X}\,[\mathbf{Sub}^1\,[\mathbf{Triv}^{(*_1\iota)}\,y]\,{}^0x\,{}^0([\mathbf{Bl}\,x]_{tw})]]_{tw})]\\ [\mathbf{PrU}_{tw} =^{\iota} \mathbf{PrA}_{tw}]\end{array}}{{}^2[\mathbf{Sub}^2\,{}^0(\mathbf{PrA}_{tw})\,{}^0y\,{}^0([\mathbf{Bel}^1\,\mathbf{X}\,[\mathbf{Sub}^1\,[\mathbf{Triv}^{(*_1\iota)}\,y]\,{}^0x\,{}^0([\mathbf{Bl}\,x]_{tw})]]_{tw})]}$$

a.-*case*.

The *de dicto* reading of the above argument consists of the following constructions C_1, C_2, C_3:

Example 56 (Substitution in belief *de dicto*)

$$\frac{\begin{array}{c}[\mathbf{Bel}^1\,\mathbf{X}\,{}^0([\mathbf{Bl}\,\mathbf{PrU}_{tw}]_{tw})]_{tw}\\ [\mathbf{PrU}_{tw} =^{\iota} \mathbf{PrA}_{tw}]\end{array}}{[\mathbf{Bel}^1\,\mathbf{X}\,{}^0([\mathbf{Bl}\,\mathbf{PrA}_{tw}]_{tw})]_{tw}}\ \not{\text{(SI)}}$$

However, C_1, C_2, C_3 are not v-congruent (in the appropriate order) with the following constructions C_1', C_2, C_3', which form an instance of (SI):

$$\text{(SI)}\ \frac{\begin{array}{c}{}^2[\mathbf{Sub}^2\,{}^0(\mathbf{PrU}_{tw})\,{}^0x\,{}^0([\mathbf{Bel}^1\,\mathbf{X}\,{}^0([\mathbf{Bl}\,x]_{tw})]_{tw})]\\ [\mathbf{PrU}_{tw} =^{\iota} \mathbf{PrA}_{tw}]\end{array}}{{}^2[\mathbf{Sub}^2\,{}^0(\mathbf{PrA}_{tw})\,{}^0x\,{}^0([\mathbf{Bel}^1\,\mathbf{X}\,{}^0([\mathbf{Bl}\,x]_{tw})]_{tw})]}$$

since the variable x for which one wants to substitute is in the scope of 0-execution, and so it has a lower order than the construction to which one wants to substitute, hence, it is not substitutable in $[\mathbf{Bel}^1\,\mathbf{X}\,{}^0([\mathbf{Bl}\,x]_{tw})]_{tw}$, cf.

8 $\text{T}\textsc{riv}^{(*_2\iota)}$ takes an individual to its 0-execution, whose lowest possible order is 1, but this 0-execution is treated here as a 2nd-order construction (cf. 2.4.4). I omit "$\cong o$" and "Γ", similarly below.

5.4.2. Therefore, this instance of (SI) does not license the above argument on the *de dicto* reading of its major premise and conclusion – which is correct.

One may notice that, since such a substitution is dull, (SI) only licenses the following argument that is not based on a real substitution (the argument is even licensed by (RM), 3.3):

Example 57 (Valid argument involving dull substitution in belief *de dicto*)

$$\frac{\text{“Xenia believes that the President of the USA is blue-eyed.”} \quad \text{“The President of the USA is the first African-American president of the USA.”}}{\text{“Xenia believes that the President of the USA is blue-eyed.”}}\ \text{(SI)}$$

$$\frac{[\mathbf{Bel}^1\,\mathbf{X}\,^0([\mathbf{Bl}\,\mathbf{PrU}_{tw}]_{tw})]_{tw} \quad [\mathbf{PrU}_{tw} =^{\iota} \mathbf{PrA}_{tw}]}{[\mathbf{Bel}^1\,\mathbf{X}\,^0([\mathbf{Bl}\,\mathbf{PrU}_{tw}]_{tw})]_{tw}}\ \text{(SI)}$$

5.4.4 Genuine/pseudo- hyperintensional context

The reader has surely noted that substitution in belief *de re* is possible, i.e., one may substitute in a hyperintensional context without losing v-congruence. This fact suggests that these particular contexts are not genuine hyperintensional contexts, for they have been defined 1.3 as resisting such substitution.

Definition 81 (Genuine/pseudo- hyperintensional contexts and substitution)

Not every syntactical occurrence of a hyperintensional operator creates a genuine hyperintensional context for its full scope. Unlike *genuine hyperintensional contexts*, proper substitution in *pseudo-hyperintensional contexts* is possible.

5.4.5 Quantification into belief contexts

The fact that a belief operator need not create a hyperintensional context for its full scope will be illustrated in an example which harbours both genuine and pseudo-hyperintensional contexts. The example is a variant of substitution, and also quantification, into belief contexts.

Example 58 (Valid argument involving quantification into belief context)

$$\frac{\text{“Wiles knows that Fermat believes FLT.”}}{\text{“There is a construction such that Wiles knows that Fermat believes it.”}}\ \text{(EG)}$$

If considered as valid, the premiss must be read in the *de re* way in which Wiles is ascribed to know, concerning FLT, that Fermat believes it, and so there is a certain construction that is believed by Fermat. The speaker says something about FLT, which is thus in the speaker's perspective, and one can therefore substitute a certain co-referring expression for “FLT”.

On the other hand, the verb “to know” otherwise preserves its role of ‘opacity operator’, for one cannot ‘substitute’ (say) “the most renowned French mathematician who was a lawyer” for “Fermat” in the subordinate clause. Briefly, “to know” creates a genuine hyperintensional context, however, not for its full scope “that Fermat believes FLT/it”.

To simplify our further considerations, let

“*FLT*” be short for “$[\forall^{\rho}\lambda x\forall^{\rho}\lambda y\forall^{\rho}\lambda z\forall^{\rho}\lambda n[[[x^n + y^n] =^{\rho} z^n] \rightarrow [n < 3]]]$”.

There are three admissible *de re* readings of the conclusion of the above argument. On each reading, the speaker claims that there is a certain 1st-order construction that is the object of Fermat's belief. The readings differ in the way they reach *FLT*. Let ${}^{(0)}\mathbf{W}$ (Wiles)/ι; ${}^{(0)}\mathbf{K}^2/(\pi\iota*_2)$;${}^{(0)}\exists^{*_1}/(o(o*_1))$; ${}^{(0)}\exists^{*_2}/(o(o*_2))$; $c^1, c'^1/*_1$; $d^2/*_2$; ${}^{(0)}\mathbf{Triv}^{(*_2*_1)}/(*_2*_1)$:

Example 59 (Quantification into belief contexts)

“There is a construction such that Wiles knows that Fermat believes it.”

1st reading	$[\exists^{*_2}[\lambda d^2[\mathbf{K}^2\ \mathbf{W}\ [\mathbf{Sub}^2\ d^2\ {}^0c^1\ {}^0([\mathbf{Bel}^1\ \mathbf{F}\ c^1]_{tw})]]_{tw}]]$
2nd reading	$[\exists^{*_1}[\lambda c'^1[\mathbf{K}^2\ \mathbf{W}\ [\mathbf{Sub}^2\ c'^1\ {}^0c^1\ {}^0([\mathbf{Bel}^1\ \mathbf{F}\ c^1]_{tw})]]_{tw}]]$
3rd reading	$[\exists^{*_1}[\lambda c'^1[\mathbf{K}^2\ \mathbf{W}\ [\mathbf{Sub}^2\ [\mathbf{Triv}^{(*_2*_1)}\ c'^1]\ {}^0c^1\ {}^0([\mathbf{Bel}^1\ \mathbf{F}\ c^1]_{tw})]]_{tw}]]$

Remarks. The difference between the 1st and the 2nd reading: d^2 ranges over the type of 2nd-order constructions, while c^1 ranges over the type of merely 1st-order ones, which is narrower; *FLT* occurs in both types. The difference between the 2nd and the 3rd reading: the latter one involves the function $\text{Triv}^{(*_2*_1)}$, which has no representation on the linguistic level, the other readings thus provide more adequate analyses.[9] Note that the 2nd-order construction c^1 occurring within round brackets is not a direct subconstruction of all these 3rd-order constructions, and so it is not substitutable – unlike the variables d^2 and c'^1, whose order is 3. The knowledge operator thus does not create a genuine hyperintensional context covering d^2 or c'^1.

Using the first reading of the above argument's conclusion and the *de re* reading of its premiss, we get:

Example 60 (Existential generalisation (EG) and belief contexts)

$$\frac{[\mathbf{K}^2\ \mathbf{W}\ [\mathbf{Sub}^2\ {}^0FLT\ {}^0c^1\ {}^0([\mathbf{Bel}^1\ \mathbf{F}\ c^1]_{tw})]]_{tw}}{[\exists^{*_2}[\lambda d^2\ [\mathbf{K}^2\ \mathbf{W}\ [\mathbf{Sub}^2\ d^2\ {}^0c^1\ {}^0([\mathbf{Bel}^1\ \mathbf{F}\ c^1]_{tw})]]_{tw}]]}\ (\mathrm{EG})$$

The argument is licensed by (EG), since the constructions are v-congruent, in the appropriate order, with the following constructions which are their logical forms (for simplicity only partial logical forms), while the logical forms present an instance of (EG) (where $d'^2/*_2;{}^{(0)}\ \mathbf{Sub}^3/(*_3\ *_3\ *_3*_3)$):

$$(\mathrm{EG})\ \frac{{}^2[\mathbf{Sub}^3\ {}^0FLT\ {}^0d'^2\ {}^0([\mathbf{K}^2\ \mathbf{W}\ [\mathbf{Sub}^2\ d'^2\ {}^0c^1\ {}^0([\mathbf{Bel}^1\ \mathbf{F}\ c^1]_{tw})]]_{tw})]}{[\exists^{*_2}[\lambda d^2\ [\mathbf{K}^2\ \mathbf{W}\ [\mathbf{Sub}^2\ d^2\ {}^0c^1\ {}^0([\mathbf{Bel}^1\ \mathbf{F}\ c^1]_{tw})]]_{tw}]]}$$

5.5 Puzzling substitutions in belief contexts

The next few (sub)sections focus on some intriguing cases of substitution into belief sentences, most of them have been frequently discussed in literature.

5.5.1 Belief as a multiple relation

Russell (e.g. [357, 445]) famously abandoned the analysis of belief attitudes as attitudes towards 'propositions' and proposed his *multiple-relation theory of belief* (he called it "multiple-relation theory of judgement"). For example, Othello's belief that Desdemona loves Cassio is not explicated as Othello's relation towards a 'proposition', but rather as Othello's relation towards Desdemona, Cassio and a 'love' relation, which form a certain 'complex'.

The theory was not adopted in the 20th century probably for the reason that Russell's 'propositional functions', which seem to be needed for its elaboration, were abandoned. Since the Russellian structured 'propositions' have recently been rehabilitated (cf. chapter 1), variants of the multiple-relation theory of belief have been proposed, see e.g. Moltmann [256].

---

9 The third reading seems to fall under the Rule R3 by Duží and Jespersen [107], who do not, however, employed Tichý's substitution function, (EG), or his deduction system.

Remember also Quine's [315] observation that, apart from the dyadic notion of belief, there also exists a notion of *triadic belief*,[10] which is evidently mentioned in sentences such as

"Ralph is such that Xenia believes [of him] that he is a spy."

Tryadic belief strongly resembles the multiple-relation theory of belief.

In writings reacting to Quine, however, triadic belief was simply understood as dyadic belief *de re*, since in both of them, a term such as "Ralph" is substitutable by co-referential expressions such as e.g. "the man in the brown hat".

Since I have explicit substitution at my disposal, I can explicitly define the notion of triadic belief in terms of dyadic belief *de re* (where $y'/\iota; {}^{(0)}\mathbf{Bel3}^1/(\pi\iota\iota*_1)$):

Definition 82 (Triadic belief)

$$\models [\mathbf{Bel3}^1\, x\, y\, c^1]_{tw} \Leftrightarrow_o [\mathbf{Bel}^2\, x\, [\mathbf{Sub}^2\, [\mathbf{Triv}^{(*_1\iota)}\, y]\, {}^0y'\, [\mathbf{Sub}^2\, c^1\, {}^0c'^1\, {}^0([c'^1\, y']_{tw})]]]_{tw}$$

Remark. The value of c^1 can be an arbitrary construction, not necessarily a construction of an 'attribute' ascribable to the value of y. To restrict the range to constructions of conditions as 'attributes', the definition should be modified appropriately.

5.5.2 Nested beliefs

In THL, analyses of sentences with *nested beliefs* are possible. Since TTT prevents circular specification of constructions, no construction can have itself in its scope and no construction can quantify over a type to which it belongs. For example, the schematic sentence of form

"...$Bel_a(Bel_a(Bel_a\, C))$",

(where "*Bel*" is a verb such as "to believe", "*a*" a singular term standing for an agent, not necessarily one and the same through the whole formula, and "*C*" a term standing for the content of the agent's belief) cannot adequately be analysed using an iterated belief operator ${}^{(0)}\mathbf{Bel}^k$.

Each nested belief, i.e. each embedded 'proposition', must be of a lower order than the superordinate 'proposition', and so the schematic sentence has to be analysed in conformity with the following scheme

...$Bel_a^m(Bel_a^l(Bel_a^k\, C))$

[10] For discussion, see Crawford [82]

where $(k+1) \leqslant l$, $(l+1) \leqslant m$, and k is the lowest possible order of the belief C (which is usually determined by the context of the sentence).

Here is a simple example (where ${}^{(0)}\mathbf{Z}$ (Zoë)$/\iota$;${}^{(0)}\,\mathbf{Bel}^2/(\pi\iota{*_2})$):

Example 61 (Nested belief)

"Zoë believes that Xenia believes that Yannis is the President of the USA."

$$[\mathbf{Bel}^2\,\mathbf{Z}\,{}^0([\mathbf{Bel}^1\,\mathbf{X}\,{}^0([\mathbf{Y} =^{\iota} \mathbf{PrU}_{tw}])]_{tw})]_{tw}$$

Remark. Here I assume the lowest possible orders: order 1 for the object of Xenia's belief C, order 2 for the object of the nested (Zoë's) belief $Bel^1_x(C)$, and order 3 for $Bel^2_z(Bel^1_x(C))$ (where "z" denotes Zoë). But it need not be so. In some context, the believer may believe^k various (say) 1st-order constructions, for $k > 1$, because she believes 1st-order constructions that are unattainable for her within order 1 (to illustrate, she derives them via some higher-order proof within which the constructions happen to occur as higher-order ones).

Substitution to nested belief sentences is naturally somewhat complicated, since one must employ nested substitutions (3.4.3). To illustrate, let us formalise the *de re* reading of the above example (I also show its fitting paraphrase), so the President of the USA now belongs to the perspective of the speaker.

Example 62 (Substitution to nested belief)

"Zoë believes of the President of the USA
that Xenia believes that Yannis is him." |

$$[\mathbf{Bel}^2\,\mathbf{Z}\,[\mathbf{Sub}^2\,[\mathbf{Triv}^{(*_1\iota)}\,\mathbf{PrU}_{tw}]\,{}^0x\;{}^0([\mathbf{Bel}^1\,\mathbf{X}\,[\mathbf{Sub}^1\,[\mathbf{Triv}^{(*_1\iota)}\,x]\,{}^0y\,{}^0([\mathbf{Y} =^{\iota} y]_{tw})]]_{tw})]]_{tw}$$

5.5.3 Indirect belief reports and cross-reference

The object of a *de dicto* or a *de re* belief attitude can be reported in an indirect way, e.g. (where ${}^{(0)}\imath^{*_1}/(*_1(o*_1))$ – the singularisation function, 2.5.1)

Example 63 (Indirectly reported belief *de dicto* and *de re*)

"Xenia believes what is believed by Yannis." |

$$[\mathbf{Bel}^1\,\mathbf{X}\,[\imath^{*_1}[\lambda c^1[\mathbf{Bel}^1\,\mathbf{Y}\,c^1]_{tw}]]]_{tw}$$

"Xenia believes of the President of the USA what is believed by Yannis." |

$$[\mathbf{Bel}^1\,\mathbf{X}\,[\mathbf{Sub}^1\,[\mathbf{Triv}^{(*_1\iota)}\,\mathbf{PrU}_{tw}]\,{}^0x\,[\imath^{*_1}[\lambda c^1[\mathbf{Bel}^1\,\mathbf{Y}\,c^1]_{tw}]]]]_{tw}$$

Remarks. The first sentence is true if Xenia really believes the construction C (if any) that is believed by Yannis. The second sentence is true in a narrower set of circumstances, namely when C is believed both by Xenia and Yannis, and the 0-execution of the individual (if any) who is the current U.S. president can be substituted in C.

The technique of substitution even enables explicit *cross-reference* to individuals *across belief attitudes*. Perhaps the most discussed example is Geach's [151] puzzle concerning *intentional identity*, which relies on various readings of the next sentence. I offer here only one of its reading.

Let $u, u'/\iota_{\omega\rho}$; $^{(0)}\wedge/(ooo)$; $^{(0)}\exists^{\iota_{\omega\rho}}/(o(o\iota_{\omega\rho}))$; $^{(0)}\mathbf{H}$ (Hob), $^{(0)}\mathbf{N}$ (Nob) $/\iota$; $^{(0)}\mathbf{Was}/(\pi\pi)$ – past tense operator (see 4.6); $^{(0)}\mathbf{Blight}$, $^{(0)}\mathbf{Kill}/(\pi\iota\iota)$; $^{(0)}\mathbf{Think}^1$, $^{(0)}\mathbf{Wonder}^1/(\pi\iota *_1)$; $^{(0)}\mathbf{Wi}$ (be a witch) $/(\pi\iota)$; $^{(0)}\mathbf{Bm}$ (Bob's mare), $^{(0)}\mathbf{Cs}$ (Cob's sow) $/\iota_{\omega\rho}$ – which is a simplification in both cases; $^{(0)}\mathbf{Triv}^{(*_1\iota_{\omega\rho})}/(*_1\iota_{\omega\rho})$.

Example 64 (Cross-reference within belief *de re*)

"Hob thinks that a witch blighted Bob's mare, and Nob wonders whether she [the same witch] killed Cob's sow." |

$$[\exists^{\iota_{\omega\rho}}[\lambda u[[\mathbf{Wi}\ [\mathbf{OccOf}^{\iota_{\omega\rho}}_{tw}\ u]]$$
$$\wedge[\mathbf{Think}^1\ \mathbf{H}\ [\mathbf{Sub}^1\ [\mathbf{Triv}^{(*_1\iota_{\omega\rho})}\ u]\ {}^0u'\ {}^0([\mathbf{Was}\ [\mathbf{Blight}\ u'_{tw}\ \mathbf{Bm}_{tw}]]_{tw})]]_{tw}$$
$$\wedge[\mathbf{Wonder}^1\ \mathbf{N}\ [\mathbf{Sub}^1\ [\mathbf{Triv}^{(*_1\iota_{\omega\rho})}\ u]\ {}^0u'\ {}^0([\mathbf{Was}\ [\mathbf{Kill}\ u'_{tw}\ \mathbf{Cs}_{tw}]]_{tw})]]_{tw}]]]$$

Another well-known puzzle was investigated by Castañeda in a series of papers (e.g. [60]), in which he noticed that, when ascribing an attitude towards 'proposition' P to a certain third person Y, it is possible that the subject of P is Y himself. In such a case, one prononominalises the term referring to Y in favour of "he (himself)" (abbreviated by Castañeda to "he*"):

"The editor of Soul knows that he (himself) is a millionaire."

The sentence has two readings. It can be read a. as a result of pronominalisation of "the editor of Soul" by "he (himself)". The editor of Soul, say Yannis, is thus ascribed believing the 'proposition' that Yannis is a millionaire.

But the sentence can also be read in another way than as a result of pronominalisation, if the speaker intended to report Yannis' belief towards the 'proposition' that the editor of Soul (whoever he is) is a millionaire. On this b.-reading, Yannis need not know that he was appointed the editor of Soul on some secret session of the journal's editorial board, or that he just inherited a considerable wealth.[11]

[11] Castañeda also illustrated such a possible reading on the example of a war hero afflicted by amnesia who forgot that he himself was the war hero wounded 100 times, and who just started to write the hero's authoritative biography.

Obviously, neither of the two readings entails the other. This is confirmed by the following constructions which demonstrate *de*pronominalisation, i.e. a correct replacement of "he (himself)" by means of "the editor of Soul". Let $^0\mathbf{Es}$ (the editor of Soul) $/\iota_{\omega\rho}$ – which is a simplification; $^0\mathbf{Mi}$ (be a millionaire) $/(\pi\iota)$; $^0\mathbf{K}^1$ (know) $/(\pi\iota *_1)$.

Example 65 (Depronominalisation in belief sentence)

"The editor of Soul knows that he (himself) is a millionaire."

a. $[\mathbf{K}^1\, \mathbf{Es}_{tw}\, [\mathbf{Sub}^1\, [\mathbf{Triv}^{(*_1\iota)}\, \mathbf{Es}_{tw}]\, {}^0x\, {}^0([\mathbf{Mi}\, x]_{tw})]]_{tw}$

b. $[\mathbf{K}^1\, \mathbf{Es}_{tw}\, [\mathbf{Sub}^1\, {}^0(\mathbf{Es}_{tw})\, {}^0x\, {}^0([\mathbf{Mi}\, x]_{tw})]]_{tw}$

THL, and TTT in general, thus offers a mighty tool for natural language analysis that was not accessible to Castañeda's contemporaries. Castañeda himself found a satisfiable formalism in the English language enriched by a somewhat complicated system of arrows indicating admissible substitutions. Instead, THL offers exact logical tools, namely SUB^k, TRIV^k and 0-execution.

Chapter 6

Limits of belief and knowledge

6.1 Overview of the chapter

In the previous chapter 5, I offered an analysis of belief sentences as reporting attitudes towards constructions of truth values, i.e. towards *o*-constructions. This conception leads to two important questions: i. whether the analysis is formally correct and so resistant to any substantial objection, ii. what logic utilises such an analysis. The present chapter treats both of these issues; despite their relative independence, they both display limited epistemic abilities of agents.

The first part of this chapter attempts to answer question i. The previous chapter contains the THL analysis of belief attitudes as *explicit belief*: from a sentence reporting such a belief attitude to a certain construction C, one cannot derive a sentence reporting an agent's attitude towards another, albeit v-congruent, construction C'. Therefore, the agent seems to be treated as a so-called *logical idiot*, as in other explicit approaches. This way the *Paradox of Logical Omniscience* (*LOP*) is avoided, see 6.2.

It is also well known that the analysis of belief attitudes in the spirit of *PWS* cannot avoid the paradox, since it allows us to infer a sentence that ascribes to an agent an attitude to whatever logical consequence of her initial belief; the agent is thus treated as a so-called *logical genius*. Various writers have tried to find a way between the two aforementioned undesirable extremes of Scylla and Charybdis.

At first sight, thus, the THL analysis of belief might seem too restrictive. In sections 6.2.1 – 6.2.3, I will attempt to demonstrate that the analysis only seems wrong due to its mistaken understanding and use. On its adequate use, it does enable the ascriptions of agent's attitudes towards further 'propositions' that are derivable from the agent's belief base.

I call the corresponding notion *derivable belief w.r.t. derivation system* (6.2.1). A derivation system aptly models an agent's limits in *inference resources*. Derivable beliefs form an agent's set of *implicit beliefs*. Derivation systems also contain *derivation rules* used by agents to infer constructions from other ones. (The notion of a derivation system is more general than that of axiomatic theory.) Thus, the approach brings an explanation how an agent's set of 'initial beliefs' relates to the set of her 'inferable beliefs' – *via* derivation systems she masters.

With regards to question ii., I repeatedly focus on THL's capability to deal with the area of *EL*. Since THL is a highly expressive system, it is possible to reconstruct various systems of EL in it. And since THL is ready to combine *epistemic modalities* with other modalities, it can be understood as *multimodal logic*. The purpose of sections 6.3.2 and 6.4 is a sample demonstration of such a translation and also a description of some of its important advantages, as well as some constraints put on it.

6.2 PWS, THL and the Paradox of Omniscience

As discussed in 1, especially the arguments about attitudes towards mathematical facts such as $1 + 2 = 3$ or FLT bring about an important complication for PWS analysis of belief sentences because the corresponding mathematical sentences denote the same proposition Verum, and one is thus allowed to 'substitute' one sentence denoting the proposition for an equivalent sentence. To illustrate, the following intuitively invalid argument is wrongly evaluated by PWS as valid:

Example 66 (Invalid argument unblocked by PWS)

$$\frac{\begin{array}{c}\text{“Fermat believed that } 1+2=3.\text{”}\\ \text{“}(1+2=3) \;=\; \forall x\forall y\forall z\forall n((x^n+y^n=z^n)\rightarrow(n<3)).\text{”}\end{array}}{\text{“Fermat believed that } \forall x\forall y\forall z\forall n((x^n+y^n=z^n)\rightarrow(n<3)).\text{”}}$$

Since Hintikka's (e.g. [175]) announcement of the problem, LOP has been intensively studied. The following formulas LO1–LO6 have been identified as *conditions for logical omniscience* (see e.g. Meyer [251], Égré [113]):

Definition 83 (Conditions for logical omniscience (LO))

LO1	$\models K_x\varphi \wedge K_x(\varphi \rightarrow \phi) \rightarrow K_x\phi$	closure of knowledge w.r.t. material conditional
LO2	$\models \varphi \Rightarrow \models K_x\varphi$	knowledge of all valid formulas
LO3	$\models (\varphi \rightarrow \phi) \Rightarrow \models (K_x\varphi \rightarrow K_x\phi)$	closure of knowledge w.r.t. to logical consequence
LO4	$\models (\varphi \leftrightarrow \phi) \Rightarrow \models (K_x\varphi \leftrightarrow K_x\phi)$	closure of knowledge w.r.t. equivalence
LO5	$\models (K_x\varphi \vee K_x\phi) \rightarrow K_x(\varphi \vee \phi)$	

LO6	$\models K_x\varphi \rightarrow K_x(\varphi \vee \phi)$	weakening of knowledge
LO7	$\models \neg(K_x\varphi \wedge K_x\neg\varphi)$	consistence of knowledge

It is easy to check that PWS is committed to LO1–LO7, and so agents are modelled as having excessive epistemic abilities. Since LO1–LO7 hold in *standard EL with PWS (SEL)*, various proposals, some of them mentioned below (a.–d.), which have been propounded to tackle LOP, get away from the framework of SEL.[1] On the other hand, some writers, e.g. Stalnaker [387], consider EL to be a normative (not descriptive) project that treats ideal (not ordinary) believers; such authors do not seek to reform SEL.

a. Among the best-known approaches solving LOP within EL, one can find the *impossible possible worlds approach* pioneered by Hintikka [175] and technically elaborated by Rantala [344, 343] using "urn models", see also e.g. Wansing [442]. As mentioned in chapter 1, impossible possible worlds (in which e.g. $1 + 2 \neq 3$), extending the domain of 'normal' possible worlds, are used to model an agent's ignorance. However, the approach provoked various philosophical objections. For recent defence of the approach, see e.g. Berto [39], Jago [188]. In THL, there is no need to admit impossible possible worlds: i. beliefs are not modelled as propositions and ii. illogical beliefs are modelled differently (5.3).

b. An appreciated approach to LO is offered by a family of *approaches based on awareness*. See esp. Fagin and Halpern [119], Meyer and Hoek [253]; for an overview of recent investigation, see Schipper [362]. As hinted in its name, a belief operator only relates an agent to the 'propositions' (represented by sets of formulas) she is aware. Undesirable beliefs are filtered out e.g. by a certain 'sieve' on models, or the belief operator is limited by means of *Montague-Scott neighbourhood semantics* (for recent work, see e.g. Benthem and Pacuit [37], Velázquez-Quesada [439], Pacuit [284]). Despite criticism e.g. by Konolige [217], the awareness approach is still popular, since the crucial idea is intuitively sound.

c. Another approach has often been called the *syntactic approach* because the objects of beliefs are formulas, for which the approach has usually been dismissed. Recently, the approach has been revived under the name *rule-based approach*, since Eberle [111], who proposed its first version, and Konolige [216], who wrote a whole monograph, had already implemented the idea that an agent derives new beliefs from the set of initial ones by means of derivation rules. See Pucella [311], Alechina, Jago, Logan [3], or Duží, Jespersen and Müller [109] for recent contributions.[2] Admittedly, investigation of knowledge and logical omniscience within *justification logic*, in which one studies reasoning steps in

[1] A comparison of approaches a.–d. with respect to their applicability in practical situations can be found in Halpern and Pucella [166].

[2] Among such approaches one can also perhaps subsume approaches investigating the block-

learning a formula, see e.g. Artemov and Fitting [12], Artemov and Nogina [15], Artemov and Kuznets [14] or Baltag, Renne and Smets [22], aligns with this line of research. My proposal offered below is rule-based but not syntactical.

d. Recently, we have been uncovering attempts to impose restrictions on an agent's ability to know formulas while implementing the idea that agents are limited in their computational power, and so are resource-bounded. In these '*algorithmic approaches*', see e.g. Halpern, Moses and Vardi [165], Duc [100], Pucella [311], Artemov and Kuznets [13] and Bílková, Greco, Palmigiano, Tzimoulis and Wijnberg [42], models are enriched by the domain of algorithms that help to order formulas according to the difficulty of their verification/learning. Artemov and Kuznets [14], for example, show how to combine this approach with the rule-based approach.

I close this section with the reminder that one of the main reasons for the analysis of a belief as an explicit belief by means of hyperintensional semantics of THL is just the possibility to block invalid inferences which are incorporated in LO1, LO3–LO6. The THL analysis of *de dicto* beliefs 5.3 makes LO1, LO3–LO6 invalid, for substitution allowing belief C_1 to be exchanged in favour of C_2 is not allowed, regardless of C_2's v-congruence to C_1 (5.4.3). LO2 and LO7 are also invalid, but because of general assumptions concerning the verb "to know".

6.2.1 Derived beliefs

Some writers would not agree with the THL analysis of belief because of its restrictiveness. This is the reason why I need to explain the method of my analysis in greater detail, thereby helping to avoid any such objection.

As mentioned above, my proposal ranks among the recent rule-based approaches. Below is the motivation. Ask whether the following argument is valid:

$$\frac{\text{"Xenia believes that } 1+2=3.\text{"} \quad \text{"} 1+2=3 \equiv 2+1=3.\text{"}}{\text{"Xenia believes that } 2+1=3.\text{"}}$$

One is naturally tempted to say something like "Sure, provided Xenia masters the commutativity rule for addition.". If Xenia had not mastered such a rule, it would not be justified to ascribe to her an attitude towards $2+1=3$, despite the fact that she knows that $1+2=3$. (Here, I naturally assume that Xenia can posses more beliefs at one time, though she is only capable of topically concentrating on one of them.) Such an answer treats the argument as an

ing of undesirable consequences on the level of substructural rules of inference, see e.g. Bílková, Majer and Peliš [41] or Sedlár [369].

enthymeme, i.e. an argument with a suppressed premiss. In the present case, the suppressed premiss is "Xenia masters the commutativity rule for addition.".

For another example, recall the argument:

$$\frac{\text{"Fermat knew that } 1+2=3\text{."} \quad \text{"} 1+2=3 \equiv \text{FLT."}}{\text{"Fermat knew that FLT holds."}}$$

Though Fermat surely knew the trivial mathematical fact $1 + 2 = 3$, it is doubtful that he was capable of deriving FLT from it (or, more precisely, from the mathematics of that time). In other words, it is doubtful whether he had knowledge such as that which Wiles, who indeed constructed FLT's proof, has.

Inspired by such examples, I will consider each agent to master *rules* from a certain set R consisting of rules which operate on certain *base of beliefs* PC, i.e. I use the idea of *derivation systems*, whose detailed definition occurs in Raclavský and Kuchyňka [338].

First note that objects of belief are not formulas, i.e. expressions of a particular language, but rather the entities the expressions express, i.e. meanings ('propositions'). In comparison with my approach, Konolige's [216] original approach (and many other approaches) was syntactical, since he employed couples $\langle B, R\rangle$, where B is a set of basic formulas the agent believes and R a set of rules operating on B's formulas.

Aside from this difference, there are even further differences which I expose after the following definition.

Definition 84 (Derivation system)

A *derivation system* is a triple

$$DS = \langle PC, R, Q\rangle,$$

where

- PC is a (not-necessarily consistent) set of primary o-constructions, i.e. a subset of all possible 'implicit beliefs' of an agent who masters DS,
- R is a set of (correct or incorrect) inference rules, and
- Q a set of conditions satisfiable by members of PC and R.

DS is thus an object of type $((o(\pi *_k))(o*_k)(o*_k))$.

Remark. By a triple $\langle PC, R, Q\rangle$ I mean here simply the partial binary function that assigns a certain Q only to one string consisting of PC and R, while it is undefined otherwise.

Convention 33 (Type of derivation systems, δ)

"$((o(\pi*_k))(o*_k)(o*_k))$" will be abbreviated to "δ".

Comments.

i. It is the speaker (or, when evaluating such arguments, even we) who ascribes mastery of a specific derivation system DS to a particular agent. Of course, this is mere estimation of the agent's real inference capabilities. Mastery of some DS can be changed both on the side of the agent and the side of the speaker who describes it. The change of derivation systems can be triggered by the agents themselves e.g. when they acquire a new belief or learn new rules. In general, agents can *revise* their belief.[3]

ii. My proposal offers a very comprehensive framework even in the sense that some derivation systems can become subjects of another derivation system. For example, an agent may ask which derivation system is better suited for some specific task. It is also possible to isolate certain sub-systems in an agent's derivation systems (for example, imagine an agent considering various derivation systems for propositional logic). Of course, agents may use other forms of derivation rules (e.g. the rules of simple natural deduction) and so their rules (as expressions) are only represented by the rules of the system I adopt here (3.1).

iii. In the above specification of PC I do not specify which subset of an agent's 'implicit beliefs' is chosen, but a natural choice of PC would be a set of constructions that are explicit (immediate, active, etc.) beliefs of an agent. Then, R would consist of the rules enabling the derivation of the agent's 'implicit beliefs'. Note that the advantage of rule-based proposals over their rivals consists in that they bring a clear logical link between an agent's 'initial/explicit beliefs' and her 'inferable beliefs' *via* rules the agent masters, which I generalise to derivation systems.

iv. Similarly to Konolige, I do not assume deductive closure of PC. As has already been well explored in *paraconsistent logic*,[4] two inconsistent beliefs do not need to lead to an explosion of beliefs. Moreover, PC need not form a coherent set at all, for many agents obviously do not have a coherent basis for their beliefs.

v. Further, I do not presuppose that all rules of R must be correct; a plenitude of agents use incorrect rules and we allow them in our model. Of course, if we study reasoning with K within a derivation system whose R contains ill-conceived rules governing K, the genuine factivity of K need not

---

[3] See works on *belief revision*, e.g. Gärdenfors and Rott [149], Gärdenfors [148], or *dynamic epistemic logic*, e.g. van Ditmarsch, van der Hoek and Kooi [94], Baltag and Renne [21].

[4] See e.g. Priest [306], Priest [308], Carnielli, Coniglio and Marcos [58], Tanaka, Berto, Mares and Paoli [397].

be preserved: although an agent using such a system believes that she derives truths, these are not genuine truths, for her operator K is not a genuine factivity operator. In other words, invalid rules only come into consideration in the cases of non-factive attitudes such as believing.

vi. Though one may usually assume that agents are generally capable of using simple rules (Conjunction Elimination, Modus Ponens, Modus Tollens, etc.), one cannot expect all of them to be able to carry out more sophisticated rules such as e.g. the Rule of Cut or the Substitution Rule.

vii. But I go even further, since I do not generally expect that all rules from R are applicable to all elements of PC. For an illustrative example, consider a category theorist who is incapable of 'putting two and two together' and conclude from facts known to her that her partner is unfaithful, the rules she masters as a mathematician are not applicable to other facts of her life.

viii. My approach treats the agent as 'omniscient' within her own capabilities: she knows everything she can derive from her current knowledge by using the derivation system she masters modulo her inference resources (for more, see 6.2.3). Note also that derivation systems can be utilised to enlighten the agent's reasoning both from a. static and b. dynamic points of view: a. my approach indicates what an agent believes, or potentially believes, depending on what is derivable w.r.t. her resources (i.e. PC, R, Q), b. my approach indicates what an agent can get to know (within limits of her resources) after some action is performed (I add some remarks on this at the end 6.2.3, although I do not fully elaborate the dynamics of reasoning there).

Examples. In the rest of this section, I sketch three obvious examples (i.–iii.) of software applications that demonstrate the usability of the notion of a derivation system for modelling an agent's derivation of beliefs.

a. For the first example, imagine that we are designing an application for withdrawing money from an ATM. It is natural to assume that the target group of agents, i.e. ATM users, share the same DS that results from the intersection of all their individual DSs. Agents inserting their cards into ATMs are assumed to have certain specific beliefs in their PCs of their DSs (e.g. that when the message "Please enter your PIN." appears on the display, they have to enter the four digit code that was sent to them by their bank, etc.); and they are also assumed to be capable of drawing some relatively simple inferences (e.g. "The ATM says that my bank account is empty." $\vdash_R$ "Repeated attempts of withdrawal will not produce €100.").

b. Suppose we are designing an application whose purpose is to increase sales of goods in a supermarket. The application utilises data from sensors monitoring the behaviour of customers during their shopping and also data describing the goods (price, etc.). By recognising that customer X did not choose a particular product from the offer because she did not manage to compare the products due to the quality/price ratio in a time acceptable for X (e.g. because it is rather time-consuming to read the fine prints), the application does the calculation. The customer is then offered the right product before leaving the supermarket, or a convenient new product is ordered to satisfy the customer's preferences inferred from the known data that form the base PC.

c. In my third example, consider programming a computer chess game in such a way that the application adapts to its user. First, the application must learn from the play behaviour of its user, i.e. change the basis of its beliefs PC. Moreover, it should be modest with regards to resources. For example, it should not prepare 1000 moves in advance when the beginner user does not know how to successfully open a game. Long responses of the application and even its brilliant strategy would drive the user away. That is, one should be careful concerning the choice of appropriately complicated rules for R and e.g. figures related to time demand, which are deposited in Q.

6.2.2 Ascribing derived beliefs

In the previous section, I exposed the idea that arguments ascribing a derivation of (new) belief (or a change of belief, etc.) can be understood as enthymemes. It enables them to be evaluated as valid, though they are prima facie not such. I will altogether consider three methods of confirming of such an argument's validity:

(a) As just mentioned, it can be assumed that such an argument suppresses a premiss according to which agent A is capable to derive in a certain derivation system DS this or that piece of belief (knowledge, etc.), and we make such a premiss explicit.

(b) It can be assumed that such an argument involves the predicate "to believe" that is intended to mean that A relates to a certain DS used by A for derivations of beliefs, and we make this definition of such a predicate explicit.

(c) It can be assumed that such an argument involves the predicate "believe" (which is explicitly or implicitly related to a certain DS), but this predicate should be modified by an implicit, or perhaps explicit, modifier that captures the agent's limited cognitive capabilities, and we make this definition of such a modifier explicit. For the sake of illustration, consider that we treat e.g. the predicate "know" rather as "know for sure" or "know from hearsay", "belief" rather as "strongly believes" or "erroneously believes" (or their non-verbalised variants).

Firstly, I elaborate method (a). To simplify the structure of our exemplary argument, let

- $C_1 := [[1+2]=3]$
- $C_2 := [\forall^{\rho}\lambda x[\forall^{\rho}\lambda y[\forall^{\rho}\lambda z[\forall^{\rho}\lambda n[[[x^n+y^n]=^{\rho} z^n]\rightarrow[n<3]]]]]]$

Now the agent's inferential abilities depicted by DS can be made explicit by adding the relevant premiss (where $\mathbf{A}$ (an agent)$/\iota; t/\rho; w/\omega;^{(0)}\mathbf{DS}^1/\delta;^{(0)}\mathbf{CapD}^1$ (be capable of deriving some construction from a set of construction within DS) $/(\pi\iota *_k (o*_1)\delta);^{(0)}\cong$ $/(o *_1 *_1); \{...C_1...\}/(o*_1)$ (a set containing C_1, see 3.2)):

Example 67 (Argument ascribing derivation of belief)

$$\frac{\begin{array}{c}[\mathbf{Bel}^1\ \mathbf{A}\ {}^0C_1]_{tw}\\ [{}^0C_1 \cong {}^0C_2]\\ [\mathbf{CapD}^1\ \mathbf{A}\ C_2\ \{...C_1...\}\ \mathbf{DS}^1]_{tw}\end{array}}{[\mathbf{Bel}^1\ \mathbf{A}\ {}^0C_2]_{tw}}$$

Of course, by adding the premiss that relates the agent to DS, one risks that this DS is inadequate as a model of A's reasoning. Or that it is not possible to derive C_2 from $\{...C_1...\}$ in DS, in which cases the argument is rendered invalid. In those cases, we may revise our estimation to which DS Xenia is related.

Modelling the validity of arguments concerning beliefs in method (b) can be looked upon as an explicit proof-theoretic variant of approaches often developed in literature where the authors used model-theoretic methods to revise the semantics of the operator "*Bel*". One advantage of my treatment of the arguments is that one need not forcibly make the agent's relatedness to DS explicit by a special premiss, since the information is 'sealed up' in the operator

$$Bel^{DS}$$

where ${}^{(0)}\mathbf{Bel}^{\mathbf{DS}^k}/(\pi\iota *_k)$, via an appropriate rule.

This operator is defined using the operator of explicit belief and a particular derivation system DS. Let $x/\iota; c_1, c_2/*_k;^{(0)} \subseteq /(o(o*_1)(o*_1));^{(0)}\mathbf{M}^k$ (master a derivation system) $/(\pi\iota\delta);^{(0)}\mathbf{Ar}/((o*_k)\delta)$ – the function that maps DS to the set of constructions on which DS operates and is undefined otherwise):

Example 68 (Rule governing belief operator implicitly related to particular DS

$$\frac{[\mathbf{Bel}^{\mathbf{DS}^k}\ x\, c_1]_{tw} \cong o \qquad [\mathbf{M}^k\ x\,\mathbf{DS}^k]_{tw} \cong o \qquad [\{c_1, c_2\} \subseteq [\mathbf{Ar}\,\mathbf{DS}^k]] \cong o}{[\mathbf{Bel}^{\mathbf{DS}^k}\ x\, c_2]_{tw} \cong o}$$

Remark. The rule governing the meaning of "Bel^{DS^k}" (which is shown here, for simplicity, only as a 'one-directional' rule) says that A (the value of the variable x) believes (in the sense Bel^{DS}) a certain construction C_2 (the value of c_2), provided A masters a certain DS. And, C_1 and C_2 occur in Ar, which is the set of all constructions derivable in DS (within the limits stated in Q) from PC of that DS. Note that the rules an agent masters are not explicitly listed in this rule but they are encapsulated in DS.

Thanks to this rule, the third premiss of the above argument can be omitted, since it is in fact 'sealed up' in the notion ${}^{(0)}\mathbf{Bel^{DS}}^k$ which would occur in it instead of ${}^{(0)}\mathbf{Bel}^k$. To assume this is felicitous when the speaker insists on the original form of the argument without the inserted premiss and we, therefore, read their use of the verb "to belief" as meaning $\mathbf{Bel^{DS}}^k$. A technical advantage of such a possibility consists in the fact that the logical behaviour of the operator can be read off from the logical parameters in R and Q, it need not be dug out from the model-theoretic semantics of the operator "*Bel*" of rival approaches.

Method (c) will be expanded upon within the next subsection.

6.2.3 Limits of an agent's inferential resources

Some authors implemented into their models of knowledge a certain model of various limitations of an agent's inference capabilities that are generally called *resources*,[5] the idea being that an agent reasons within the limit of its *resource pool* (*RP*).

In my above model, purely deductive resources of agent A are built into components of the DS A masters: PC contains only some constructions, set R too; Q contains various restrictions put on members of PC and R. They all express resources A possesses. Now I add another way of capturing A's resources that is also capable of effectively detecting of the agent's *beliefs derivable w.r.t. DS*.

Authors who treated inferential resources investigated various options. My approach seems to resemble '*algorithmic approaches*', see e.g. Halpern, Moses and Vardi [165], Pucella [311]. The authors enriched their models by domain of algorithms, which help to order beliefs (formulas) according to the difficulty of their verification. However, these algorithms are logically primitive objects, so they are not cognates of THL constructions.

Such ordering of beliefs can be expressed within the parameter Q too, since beliefs (or rules) can be ordered (say) by their complexity. Now I focus on

[5] Their role was stressed by Duc [99, 100], who introduced the idea of feasible knowledge operator. See also references in 6.2. The present section was written in collaboration with my colleague Pezlar, and published in Czech as Raclavský and Pezlar [340], while our colleague Kuchyňka reminded us of the general idea.

an additional method of resource control by means of the *belief modifier* F^k ('filter'). F^k is a function that maps each belief of type $(\pi\iota*_k)$ to its kindred $(\pi\iota*_k)$-object.

Let us illustrate F^k and its usefulness on the example in which one models inferences made by players of a tarot card game. Suppose that in the auction phase of a particular play, one of the players, A, evaluates the strength of her hand so that she announces bidding B, meaning she will be able to meet a particular contract, say 'Garde'. The other players may believe B or not, yet what is crucial here is that they do not speculate too much about A's reasoning leading to B (i.e., what is currently in A's belief set, etc.), they rather reassess A's believing B using a belief modifier F^k. F^k epitomises A's decreasing capability to implement a winning strategy after a few near tricks due to her limited concentration, limited memory, etc. (It would often be impractical to put all these factors F^k into Q of a DS, though it is, of course, possible.)

For a simple example illustrating a technical implementation, I will consider resources to be simply numbers. These are the arguments for the function that represents *inference costs* (CI) for the derivation of a construction from a given basis of beliefs PC. For a simple illustration,[6] one can imagine the evaluation of inferences performed by students in their introductory logic courses, when e.g. an application of De Morgan's rule receives the value of three points if the input formula has three binary connectives. The complexity of inferences depends on complexity of constructions and a number and difficulty of steps. Since complexity is also represented by numbers, the cost of a particular inference is represented by substracting a corresponding number from their resource pool.

The modifier F^k can then be defined as follows (the definition is simplified to a 'one-directional' rule). Let ${}^{(0)}\mathbf{F}^k/((\pi\iota*_k)(\pi\iota*_k))$;${}^{(0)}1/\rho$;${}^{(0)}-$,${}^{(0)}+/(\rho\rho\rho)$; ${}^{(0)}< /(o\rho\rho)$; ${}^{(0)}\mathbf{IR}/(\rho\iota)_{\omega\rho}$;[7] $g/*_k$ (variable for possibly empty sets Γ of o-constructions); d/δ (variable for derivation systems DS); ${}^{(0)}\Rightarrow_{DS} /(o(o*_k)*_k\delta)$ (logical consequence relative to a derivation system); ${}^{(0)}\mathbf{CI}/(\rho(o*_n)*_n\delta)$;[8] ${}^{(0)}\forall^{*_k}/(o(o*_k))$.

[6] Indeed a very simple one. It is more standard to measure the size of formulas and proofs, see e.g. Pudlák [312] and e.g. Artemov and Kuznets [14] for elaboration.

[7] The inferential resource function IR takes an agent and returns the number of her inferential resources given a particular time and world.

[8] The function CI evaluates the cost of inferring C from Γ in some DS. The cost of an inference will also depend on the used DS, since some theorems are easier to prove in one system than another.

Example 69 (Rule for a belief operator modifier)

$$\frac{\begin{array}{c}\Big[\big[[\forall^{*_1}[\lambda c_1[[g\ c_1] \rightarrow [\mathbf{Bel}^k\ x\ c_1]_{tw}]]] \wedge [\Rightarrow_{DS}\ \ g\ c_2\ d] \wedge [\mathbf{M}^k\ x\ d]_{tw} \\ \wedge[[\mathbf{CI}\ g\ c_2\ d] < [\mathbf{IR}\ x]_{tw}]\big] \rightarrow \big[[\mathbf{Bel}^k\ x\ c_2]_{tw} \wedge [[\mathbf{IR}\ i]_{tw} - [[\mathbf{CI}\ g\ c_2\ d] + 1]]\big]\Big] \cong o\end{array}}{[[\mathbf{F}^k\ \mathbf{Bel}^k]\ x\ c_2]_{tw} \cong o}$$

Remark. Informally, F^k checks whether agent A (the value of x), who believes C_1 (the value of c_1), knows assumptions Γ (the value of g), and the construction C_2 (the value of c_2) is derivable from Γ in the given derivation system DS (the value of d) he masters, and A has enough inferential resources to derive C_2. If so, we can conclude that A knows C_2 and decreases her inferential resource pool accordingly.

Many researchers also explored the role of time in the inferential practices of agents (e.g., duration of beliefs and temporal limitation of inferences). For example, Drapkin and Perlis [118], Elgot-Drapkin, Miller and Perlis [117] proposed to take into account the stepwise nature of inference, which then led into the development of their step-logic; see also Smets and Solaki [375]. In THL, which explicitly works with a time parameter, one could achieve similar results e.g. by modifying the definition of F^k. The results of dynamic epistemic logic, e.g. its discrimination between *Bel* and *Sbel* (strong belief), or defeasible knowledge (see e.g. van Ditmarsch, van der Hoek and Kooi [94], Baltag, Renne and Smets [22], Rasmussen [345]) can also be adopted to THL.

6.3 THL and EL à propositional modal logic

Before translating EL to THL, it is important to investigate the understanding of epistemic operators K and *Bel* as *modal operators* of *necessity* $\Box$ and *possibility* $\Diamond$.

If the operators operated on propositions, one would deal with a mere extension of standard modal logic. I show such an extension in the next section 6.3.2, since it also illustrates THL's handling of modal operators.

But such an EL suffers from LOP as much as standard systems of EL (cf. 6.2). If not implementing a non-standard adjustment of such systems (6.2), the systems of EL that treat objects of beliefs as 'propositions' are more suitable. I show how to build crucial items of such systems in 6.4.

6.3.1 Modal operators

In THL, modal operators can be (even simultaneously) treated in the following ways:

(a1) $\Box$ and $\Diamond$ apply to propositions,

(a2) $\Box$ and $\Diamond$ apply to propositions as modifiers,

(b) $\Box$ and $\Diamond$ apply to constructions of truth values (or propositions),

(c) $\Box$ and $\Diamond$ apply to formulas.

In this section, I deal mainly with approach (a2). Approach (b) will be discussed in 6.4; (c) in 7. Only approach (b) to modal operators provides suitable analogues of epistemic operators as I view them in this book.

Approach (a) defines modal operators either (a1) as sets of propositions,[9] i.e. as objects of type $(o\pi)$, or (a2) as propositional modifiers, i.e. as objects of type $(\pi\pi)$ (see Raclavský [322]).

Approach (a1) aligns with the conception of modal operators as quantifiers for possible worlds,[10] since they are defined in terms of them (where $p, p'/\pi$; ${}^{(0)}\forall^{\rho}, {}^{(0)}\exists^{\rho}/(o(o\rho))$; ${}^{(0)}\forall^{\omega}, {}^{(0)}\exists^{\omega}/(o(o\omega))$; ${}^{(0)}\underset{o}{\Box}, {}^{(0)}\underset{o}{\Diamond}/(o\pi)$):

Definition 85 (Modal operators as quantifiers over possible worlds and moments of time)

$\models [\underset{o}{\Box}\, p] \Leftrightarrow_o [\forall^{\rho}[\lambda t[\forall^{\omega}\lambda w.p_{tw}]]]$
$\models [\underset{o}{\Diamond}\, p] \Leftrightarrow_o [\exists^{\rho}[\lambda t[\exists^{\omega}\lambda w.p_{tw}]]]$

Modal operators defined as modifiers are convenient for natural language analysis and also for the construction of purely intensional logics (where ${}^{(0)}\underset{\pi}{\Box}, {}^{(0)}\underset{\pi}{\Diamond}/(\pi\pi)$):

Definition 86 (Modal operators as modifiers)

$\models [\underset{\pi}{\Box}\, p] \Leftrightarrow_{p'} [\lambda t \lambda w[\underset{o}{\Box}\, p]]$
$\models [\underset{\pi}{\Diamond}\, p] \Leftrightarrow_{p'} [\lambda t \lambda w[\underset{o}{\Diamond}\, p]]$

Comments.

i. *Properties of propositions.* In approach (a1), $\underset{o}{\Box}$ is the set that contains solely the *analytic proposition* which is true (i.e. it has the value T) in all Ts and Ws; I will call it *Verum*. In conception (a2), $\underset{\pi}{\Box}$ maps Verum to Verum and every other proposition to the analytic proposition *Falsum* (as I will call it) which is false in every T and W. $\underset{o}{\Diamond}$ contains all *possible propositions*, which

[9] That is, similarly as in Tichý's TIL, see e.g. Tichý [418], Raclavský, Kuchyňka and Pezlar [339] or Raclavský [335].

[10] See e.g. Montague [259], Kuhn [222], Reinhardt [346], Fitting and Mendelsohn [129], Halbach and Welch [164].

comprises all *contingent propositions* and Verum, yet it does not contain Falsum and the proposition which is constantly gappy, i.e. the proposition which I will call *Vacuum.* $\underset{\pi}{\Diamond}$ maps each possible proposition to Verum.[11]

ii. *Interdefinability of modal operators.* By adding the proposition Vacuum to Verum and Falsum, the familiar interdefinability (or: exchange) of modal operators '$\Diamond\varphi \Leftrightarrow \neg\Box\neg\varphi$' does not hold in THL. To see this, imagine, when adopting approach (a2), the appropriate constructions; then, if φ yields Vacuum, the construction on the left-hand side of $\Leftrightarrow$ is false, whereas the construction on the right-hand side of $\Leftrightarrow$ is true. To preserve the rule, one must amend partiality by means of the operator for the totalisation of proposition (where ${}^{(0)}\underset{\pi}{\neg}, {}^{(0)}\mathbf{Tot}^{\pi}/(\pi\pi)$ – see 4.4.3):

Definition 87 (Interdefinability of modal operators)

$\models [\underset{\pi}{\Diamond}\,[\mathbf{Tot}^{\pi}p]] \Leftrightarrow_{p'} [\underset{\pi}{\neg}[\underset{\pi}{\Box}[\underset{\pi}{\neg}[\mathbf{Tot}^{\pi}p]]]]$

iii. *Iteration of modal operators.* In conception (a2), iterations of modal operators well known from S5 make good sense and are easily provable (which is why I put "$\vdash$"). By omitting some couples of appropriate brackets, one has e.g.

Example 70 (Iteration of modal operators)

$\vdash [\underset{\pi}{\Box}\underset{\pi}{\Box}...\underset{\pi}{\Box}\,P] \Leftrightarrow_{p'} [\underset{\pi}{\Box}\,P]$

$\vdash [\theta\,\underset{\pi}{\Box}\,P] \Leftrightarrow_{p'} [\underset{\pi}{\Box}\,P]$

where θ is an arbitrary sequence consisting of ${}^{(0)}\underset{\pi}{\Box}$s and/or ${}^{(0)}\underset{\pi}{\Diamond}$s, and P is an arbitrary construction of a proposition, i.e. P/π.

6.3.2 A sample derivation system governing a modal operator

Now I am going to build up a sample derivation system governing $\underset{\pi}{\Box}$ that would resemble the familiar S5 system. It will differ from the textbook variant of S5 because of the employment of partiality and the temporal parameter.[12]

---

[11] For definitions of the discussed properties of propositions, see e.g. Raclavský, Kuchyňka and Pezlar [339]. I add that the just discussed modal operators can be weakened to concern solely the parameter W, or merely T.

[12] In the very recent paper by Johannesson [197], one can find partial semantics for quantified modal logic; it implements weak Kleene logic, it has a version with varying domains. The present approach is much richer, since it also involves strong Kleene logic, functional terms, and ramification.

Definition 88 (Language $\mathcal{L}_{\underset{\pi}{\Box}}$)

$$\varphi ::= P \mid (\varphi) \mid [\underset{\pi}{\neg}\varphi] \mid [\varphi \underset{\pi}{\wedge} \psi] \mid [\underset{\pi}{\Box}\varphi] \mid [\underset{\pi}{\Diamond}\varphi]$$

where

- 'propositional letters' P, P', etc., are meta-variables representing any (kth-order) variable ranging over π, i.e. $P, P'/\pi$,
- φ, ψ, etc. are meta-variables representing arbitrary π-constructions of a given order k,
- "(φ)" is in fact short for "$[\mathbf{Tot}^{\pi}\ \varphi]$",
- brackets are omitted, provided no ambiguity may arise.

Remarks. The definition can be understood as circumscribing a certain formal language. But not necessarily, for we can view it as a specification of a certain set of constructions, while the denotational semantics for the language, which occurs below, describes their v-constructing.

Definition 89 (Model for $\mathcal{L}_{\underset{\pi}{\Box}}$)

A *model* $\mathfrak{M}$ for $\mathcal{L}_{\underset{\pi}{\Box}}$ is an object of type $((\pi(*_k(\rho\omega)))(o(\rho\omega))(o(\rho\omega)(\rho\omega)))$, it is the triple

$$\mathfrak{M} = \langle W \times T, R, V \rangle,$$

where

- $W \times T$ is the universal set of $\langle W, T \rangle$ couples, i.e. an object of type $(o(\rho\omega))$,
- R is an *accessibility relation* between $\langle W, T \rangle$ couples, i.e. an object of type $(o(\rho\omega)(\rho\omega))$,
- the *valuation map* V is a binary function of type $(\pi(*_k(\rho\omega)))$ that maps every (kth-order) P to a proposition, dependently on the value of variable s (or s', etc.), which is a $\langle W, T \rangle$ couple.

Remarks. A so-called *pointed model* is a couple $\langle \mathfrak{M}, s \rangle$, where s stands for the 'current' couple $\langle W, T \rangle$. The function V should not be confused with the valuation v. I use $\langle W, T \rangle$ couples rather than $\langle T, W \rangle$ couples to underline the resemblance to standard modal logic.

Convention 34 (Type of modal model, μ)

"μ" stands for "$((\pi(*_k(\rho\omega))(o(\rho\omega)(\rho\omega))(o(\rho\omega)))$".

The proper (algorithmic) semantics of $\mathcal{L}_{\underset{\pi}{\Box}}$'s formulas is provided by the constructions displayed by the formulas. The constructions v-construct propositions that present the *denotational semantics for* $\mathcal{L}_{\underset{\pi}{\Box}}$:

Definition 90 (Denotational semantics for $\mathcal{L}_{\underset{\pi}{\Box}}$)

$$
\begin{array}{lll}
\mathfrak{M}, s \models P & \text{iff} & V(P, s) = \text{Verum} \\
\mathfrak{M}, s \models \underset{\pi}{\neg}\varphi & \text{iff} & \text{it is not true that } \mathfrak{M}, s \models \varphi \\
\mathfrak{M}, s \models (\varphi) & \text{iff} & \text{there exists a truth value of } \mathfrak{M}, s \models \varphi \text{ that equals T} \\
\mathfrak{M}, s \models \varphi \underset{\pi}{\wedge} \psi & \text{iff} & \mathfrak{M}, s \models \varphi \text{ and } \mathfrak{M}, s \models \psi \\
\mathfrak{M}, s \models \underset{\pi}{\Box}\varphi & \text{iff} & \text{for every } s' \text{ with } sRs'\text{: } \mathfrak{M}, s' \models \varphi \\
\mathfrak{M}, s \models \underset{\pi}{\Diamond}\varphi & \text{iff} & \text{for some } s' \text{ with } sRs'\text{: } \mathfrak{M}, s' \models \varphi
\end{array}
$$

Remarks. Note that the denotational values of $\mathcal{L}_{\underset{\pi}{\Box}}$'s members are propositions, not truth values. For this and several following definitions, it holds that all their parts can be reformulated from English meta-language to $\mathcal{L}_{\mathsf{TTT}}$, and so fully manipulated as logical objects within TTT. In such a reformulation, there would occur couples $\langle\mathfrak{M}, s\rangle$, which are partial functions of type $((\rho\omega)\mu)$, and $\models$ would be a certain total relation(-in-extension) of type $(o((\rho\omega)\mu))\pi)$.

Definition 91 (Validity in $\mathcal{L}_{\underset{\pi}{\Box}}$)

φ is called (logically) *valid*, written in this section "$\models \varphi$", iff $\mathfrak{M}, s \models \varphi$ for every $\mathfrak{M}$ and $\langle W, T\rangle$.

To formulate e.g. the popular Axiom of Modal Distribution in our derivation system, an extension of our language is needed:

Definition 92 (Language $\mathcal{L}'_{\underset{\pi}{\Box}}$)

$\mathcal{L}'_{\underset{\pi}{\Box}} ::= \mathcal{L}_{\underset{\pi}{\Box}}$ extended by $\varphi \underset{\pi}{\rightarrow} \psi$

Definition 93 (Derivation system $DS_{\underset{\pi}{\Box}}$)

A *derivation system* $DS_{\underset{\pi}{\Box}}$ is simply a couple

$$DS_{\underset{\pi}{\Box}} = \langle CS, R\rangle,$$

which is a partial function of type $((o*_k)(o*_k))$ that maps the set CS of (all) members of the set of construction depicted by the terms of $\mathcal{L}'_{\underset{\pi}{\Box}}$ to the set R which contains the following constructions a.–b., otherwise it is undefined. Let q/π does not occur freely in φ or ψ, and ${}^0\vdash_{DS_{\underset{\pi}{\Box}}} /(o*_{k-2})$ (for $3 \leqslant k$):

a.

$\models \emptyset \Rightarrow [\underset{\pi}{\Box}\varphi \underset{\pi}{\rightarrow} \varphi] \cong q$	*Ax. of Necessity ('T-axiom')*
$\models \emptyset \Rightarrow [\underset{\pi}{\Diamond}\varphi \underset{\pi}{\rightarrow} \underset{\pi}{\Box}\underset{\pi}{\Diamond}\varphi] \cong q$	*Ax. 5*
$\models [\underset{\pi}{\Box}[(\varphi) \underset{\pi}{\rightarrow} (\psi)] \cong q \Leftrightarrow_o [\underset{\pi}{\Box}(\varphi) \underset{\pi}{\rightarrow} \underset{\pi}{\Box}(\psi)] \cong q$	*Ax. of Modal Distribution*
$\models [\varphi \underset{\pi}{\rightarrow} \psi] \cong q, \varphi \cong q \Rightarrow \psi \cong q$	*Modus Ponens* (a modal variant)
$\models [\underset{\pi}{\neg}[(\varphi) \underset{\pi}{\rightarrow} (\psi)]] \cong q \ \Leftrightarrow_o [(\varphi) \underset{\pi}{\wedge} \underset{\pi}{\neg}(\psi)] \cong q$	*The Rule of $\underset{\pi}{\rightarrow}$-to-$\underset{\pi}{\wedge}$ Conversion*
$\models [\underset{\pi}{\neg}[\underset{\pi}{\Diamond}[\underset{\pi}{\neg}(\varphi)]]] \cong q \Leftrightarrow_o [\underset{\pi}{\Box}(\varphi)] \cong q$	*The Rule of Modal Operators Exchange*
$\models [{}^0\vdash_{DS_{\underset{\pi}{\Box}}} \varphi] \cong q \Rightarrow [{}^0\vdash_{DS_{\underset{\pi}{\Box}}} \underset{\pi}{\Box}\varphi] \cong q$	*The Rule of Necessitation*

b.
Further members of R are all 'tautologies' of partial propositional logic, whose language is $\mathcal{L}_{PL} ::= P \mid (\varphi) \mid [\underset{\pi}{\neg}\varphi] \mid [\varphi \underset{\pi}{\wedge} \psi] \mid [\varphi \underset{\pi}{\rightarrow} \psi]$.

Remarks. The totalisation of propositions is only used where appropriate, cf. the round brackets. Note that the greater the k, the greater is the expressive power of such a derivation system, for its higher-order language is more comprehensive.

6.4 Remarks on translating EL to THL

As announced above, the aim of this section is not to propose a particular system of EL, but to provide methodological remarks on translating various (existing) systems of EL to THL, and so to the framework of TTT.

1)
We have seen in the previous chapter (5.4.2) that an absorption of EL to TTT – or: an improvement of EL by utilising TTT – requires a heavy upgrade of EL's fundamental idea, according to which knowledge/belief consists in the determination of possible epistemic scenarios as sets of $\langle W, T\rangle$ couples:

- the original idea of EL, according to which knowing/believing are attitudes towards propositions, is elevated from the level of intensions to the level of constructions.

The notions of knowledge, belief, etc., are therefore explicated as determining conditions satisfiable by individuals and kth-order o-constructions, i.e. as objects of type $(\pi\iota{*}_k)$, while the conditions are v-constructed by operators such as $\mathbf{K}^k$.

2)

Recall also that our analysis of epistemic notions is framed in a certain RTT, in consequence of which

- operators of knowledge and also objects of knowledge are typed, which amounts to imposing restrictions on substitution (and so also on quantification).

For example, the construction $[\mathbf{K}^k\, x\, c^k]_{tw}$, which typically occurs in translations of EL's formulas, is of order $k+1$, and so it cannot be in the range of its variable c^k which consist of kth-order constructions. Consequently, it cannot be substituted for c^k.

The main reasons for the restrictions are:

(i) the non-circular construction of 'propositions' explicated as o-constructions,

(ii) a resistance to a family of epistemic paradoxes,

(iii) an ability to capture nuances the standard notation of EL lacks.

The reason (i) was mentioned in 2.2: TTT implements VCP. Note that the restrictions arising from this reason imitate some principal limits of an agent's epistemic capabilities. For many partial reasons supporting (i) and (iii), see the previous chapter (e.g. 5.5.2); in the rest of this section, I only introduce some unstated reasons and remarks. Point (ii) will be treated in the next chapter, 7.

The intuitive appeal for type restrictions which are displayed by orders of constructions can be documented e.g. in the sentence "Socrates knows that he knows nothing.". On its ordinary formalisation,

"$K_s(\forall\varphi\neg K_s\,\varphi)$",

where "s" denotes Socrates, the sentence is bound to sound nonsensical or paradoxical. In THL, an analysis of the nonsensical reading is possible using the v-improper construction of form $K^1_s(\forall\varphi^1\neg K^1_s\,\varphi^1)$ (this is enabled by composition that need not be type-theoretically coherent, cf. e.g. my analysis of belief attitudes towards 'nonsensical propositions' in 5.3). Yet THL is not forced to propose it as the only analysis, as in the common approach, since it has a greater expressive capability.

In THL, the statement can be formalised as a possibly true higher-order statement of form $K^2_s(\forall\varphi^1\neg K^1_s\,\varphi^1)$. (Or, on its not so much faithful reading, as $\forall\varphi^1 K^2_s(\neg K^1_s\,\varphi^1)$.) Let ${}^{(0)}\mathbf{S}$ (Socrates) $/\iota;{}^{(0)}\underset{\pi}{\neg}/(\pi\pi);c^1,c'^1/*_1;{}^{(0)}\forall^{*_1}/(o(o*_1))$; ${}^{(0)}\mathbf{K}^1/(\pi\iota*_1)$; ${}^{(0)}\mathbf{K}^2/(\pi\iota*_2);{}^{(0)}\mathbf{Sub}^2/(*_2\,*_2\,*_2*_2)$:

Example 71 (Sentence expressing someone's knowledge concerning his own knowledge)

"Socrates knows that he knows nothing." |
$[\mathbf{K}^2\,\mathbf{S}\,{}^0([\forall^{*_1}[\lambda c^1\,[\mathbf{Sub}^2\,c^1\,{}^0c'^1\,{}^0([\underset{\pi}{\neg}[\mathbf{K}^1\,\mathbf{S}\,c'^1]]_{tw})]]]])]_{tw}$

3)

The translation must also face the following two facts: (i) the range of variables for constructions, e.g. $c^k/*_k$, contains o-constructions as well as constructions of objects of other types; (ii) in many cases, one needs the value of c^k to be v-proper and v-constructing a certain truth value.

To illustrate, consider EL's variant of the T-axiom,

$$K_x\varphi\rightarrow\varphi \qquad \textit{the Axiom of Knowledge}$$

whose straightforward, but wrong, translation is the following construction (where $x/\iota;{}^{(0)}\!\!\rightarrow/(ooo)$):

$$[[\mathbf{K}^k\,x\,c^k]_{tw}\rightarrow\,{}^2c^k]$$

If its consequent did not v-construct a certain truth value, either for the reason (i), or (ii), the law would not hold, since the whole construction would be v-improper.

Both complications can effectively be removed by utilising *alethic normalisation function* $/\cdot/$ (often written as "$/C/$"), which is a certain analogue of $det(C)$ (see 2.6) that maps constructions to truth values. But $/C/$ maps constructions to o-constructions, while it amends the possible gaps in C's v-constructing by the value F, and if C is not an o-construction, it maps C to ${}^{(0)}$F.

Definition 94 (Alethic normalisation function, $/\cdot/$)

Let C,C' be constructions of order $*_k$ and ${}^{(0)}\mathtt{F}/o$; ${}^0/\cdot//(*_k*_k)$.

$$/C/=\begin{cases}C & \text{if } \forall v\exists o([\![C]\!]^{\mathfrak{M},v}=o)\\ C' & \text{if } C \text{ is an } o\text{-construction and } \exists v\neg\exists o([\![C]\!]^{\mathfrak{M},v}=o)\\ {}^{(0)}\mathtt{F} & \end{cases}$$

where C' is an o-construction such that

$$[\![C']\!]^{\mathfrak{M},v} = \begin{cases} \mathtt{T} & \text{if } [\![C]\!]^{\mathfrak{M},v} = \mathtt{T} \\ \mathtt{F} & \text{if } [\![C]\!]^{\mathfrak{M},v} = \mathtt{F}, \text{ or } \neg\exists o([\![C]\!]^{\mathfrak{M},v} = o) \end{cases}$$

Example 72 (Axiom of Knowledge)

$[[\mathbf{K}^k\, x\, /c^k/]_{tw} \rightarrow {}^2/c^k/]$

Remarks. Note that the law captures an important property of our notion of knowledge: knowledge is concerned with 'propositions', i.e. with something which is true or not. Knowledge that is not related to 'propositions' (e.g. knowledge of stars or knowledge of how to create a symphony) does not count as a type of belief which is investigated in this book. Another important remark: an axiom (as well as the corresponding rule) should be conditioned in order to avoid the destructive influence of constructions such as $\boxed{K^k}$ for the reasons discussed in the next chapter (7.5).

Instead of an axiom, $\mathrm{ND}_{\mathsf{TTT}}$ derivation systems (3.1) contain a corresponding rule:

Example 73 (The Rule of Factivity of K)

$$\frac{[\mathbf{K}^k\, x\, /c^k/]_{tw} \cong \mathtt{T}}{{}^2/c^k/ \cong \mathtt{T}}$$

Condition: Substitution of a construction executing $\boxed{K^k}$ during its v-constructing for c^k is not permitted.

Remarks. The rule can be adjusted so that the alethic normalisation function would not be present explicitly, but by an appropriate condition of the rule. In the rule, I do not allow an arbitrary truth value to be presented on the right side of $\cong$; $\mathtt{T}$ is required for the fact that not knowing C^k does not entail that C^k is false. The condition related to $\boxed{K^k}$ is discussed in 7.5 and 7.6. Since the condition is perhaps too restrictive, I would rather like to speak about vigilant treatment with constructions containing $\boxed{K^k}$.

4)

The previous chapter suggests that the crucial part of translating EL to THL is the fact that

- the translation of some EL's formulas utilises substitution 3.4 in the scope of epistemic operators.

For the sake of illustration, I render the familiar

$K_x\varphi \to K_xK_x\varphi$ *the Law of Positive Introspection*

as (where $y/\iota; d^1/*_1$; ${}^{(0)}\mathbf{Sub}^1/(*_1 *_1 *_1 *_1)$):

Example 74 (The Law of Positive Introspection)

$$[[\mathbf{K}^1\, x\, /c^1/]_{tw} \to [\mathbf{K}^2\, x\, [\mathbf{Sub}^2\, /c^1/\, {}^0d^1\, [\mathbf{Sub}^1\, [\mathbf{Triv}^{(*_1\iota)}\, x]\, {}^0y\, {}^0([\mathbf{K}^1\, y\, d^1]_{tw})]]]_{tw}]$$

Remarks. I keep the lowest possible orders here. The translation of further laws of EL is similar (but see point 5 below). In the present example, the whole construction is of order 3 and so the 3rd-order variable c^1 v-constructs certain 2nd-order constructions whose lowest possible order is 1 (cf. a similar case in 5.4.5). Note that I do not attempt to resolve here the issue whether the rules of EL are philosophically sound or not.

5)

Note that EL's translation to THL (and thus TTT) illuminates logically interesting nuances. Consider e.g.

$K_x(\varphi \wedge \psi) \vdash K_x\varphi \wedge K_x\psi$ *the Rule of Distributivity of K*

which can be read in various ways, since there is a number of possibilities how orders can be 'distributed' through the subconstructions (even if we omit cases resulting in v-improper constructions). Differences among readings are caused, among other things, by various possible orders k_i of the operator ${}^{(0)}\mathbf{K}$ in its various occurrences within the whole construction and also the differences in orders of constructions c and d.

I show the rule in its 'minimalistic' reading, on which 'propositional' variables range over 1st-order constructions (where $d'^1/*_1$):

Example 75 (The Rule of Distributivity of K over $\wedge$)

$$\frac{[\mathbf{K}^2\, x\, [\mathbf{Sub}^2\, /c^1/\, {}^0c'^1\, [\mathbf{Sub}^1\, /d^1/\, {}^0d'^1\, {}^0([c'^1 \wedge d'^1])]]]_{tw} \cong \mathrm{T}}{[[\mathbf{K}^2\, x\, /c^1/]_{tw} \wedge [\mathbf{K}^2\, x\, /d^1/]_{tw}] \cong \mathrm{T}}$$

Chapter 7

Belief and paradoxes

7.1 Overview of the chapter

An important argument in favour of my type-theoretic approach to belief, and for EL built within it, consists in its resistance to a number of renowned paradoxes. Before WWII they were called 'epistemic paradoxes', but *semantic paradoxes* (e.g. the Liar Paradox) caused by our mistaken intuitions about the semantics of language were removed from this group (Ramsey [341]), and so contemporary *epistemic paradoxes* only consist of paradoxes concerned with belief, esp. knowledge.[1]

It is well known that Russell ([353], [445] with Whitehead) demonstrated TT's capability to solve all logical and 'epistemic paradoxes' known to him due to his analysis of 'propositions' and 'propositional functions'. In this chapter, I partly confirm Russell's optimism by solving selected epistemic and semantic paradoxes within TTT.[2]

To start, I show type restrictions when solving the not-so-well-known Bouleus Paradox (7.2) and its novel mate, Brandenburger-Keisler Paradox. When examining the Preface Paradox, we find that the method is not all-encompassing, a certain additional assumption concerning the notion of belief must be either adopted or rejected.

To deepen my investigation, I also provide a solution to famous semantic paradoxes within TTT (7.3). It may perhaps be called *'hierarchical solution'*, for it partially follows the 'hierarchical approach' by Russell and Tarski. However, I repeatedly show that hierarchisation of 'propositions' (i.e. *o*-constructions)

---

[1] See the entry by Sørensen [385] for an overview of epistemic paradoxes from the current perspective and the entry by Cantini and Bruni [55] for paradoxes from the viewpoint of modern logic. Specifically for the Liar Paradox, see e.g. Beall, Glanzberg and Ripley [33], Visser [440]. See also further references below.

[2] I exclude the epistemic paradoxes that are not related to the main topics of the book, e.g. the Lottery Paradox. With regards to Russell's paradox, remember that I briefly described the solution to it both in STT and (in fact) RTT in section 2.2.1.

does not bring a universal method for solving semantic and selected epistemic paradoxes: TTT and other RTTs that admit a reducibility principle are so expressive that some paradoxes may reappear, and so an additional method of solving paradoxes must be used.

I examine limitations of explication of notions such as knowledge, belief, truth or assertion (7.4), which I will call *'propositional' notions*, which results from limitations of logical space. One of my theorems can be viewed as a paradox which I call the Paradox of Impossibility of Knowing One's Own Ignorance (7.5). Using the results, I analyse the famous Knower Paradox (7.6) that is often studied within EL and the theory of belief attitudes.

I also investigate the famous Church-Fitch Paradox (*FP*, 7.7 and further). Like the Knower Paradox, it combines both modalities and epistemic notions. Although the 'hierarchical solution' was dominant as a solution to semantic paradoxes from WWI until the mid-1970s, and is still an accepted solution e.g. to the Russell-Myhill paradox and perhaps even the Knower Paradox, its application to the FP immediately faced criticism. I show that the criticism is based on mistakes and confusions. However, I reworked the FP to a form TTT cannot solve (this does not affect the conception of meaning adopted in this book, but rather verificationism).

7.2 The Bouleus Paradox and the Preface Paradox

According to the *Bouleus Paradox* (see Church [66]), Bouleus uttered the sentence

> "I have been sometimes mistaken."

as his only statement. A vicious circular self-reference thus seems to occur.

As shown in the example of Socrates' famous statement (6.4), Bouleus' statement can schematically be formalised using the indication of orders as follows:

$$P^B: \qquad \exists t \exists p^1 (Bel^1_b(p^1, t) \wedge \neg T^1(p^1))$$

where p^1 is a variable for 1st-order 'propositions', T^1 is a 'truth-predicate' applicable to such 'propositions', Bel^1 is a belief attitude, t is a variable for moments of time, "b" denotes Bouleus. According to TTT, P^B is of order 2 and so it cannot belong to its subject matter (Church [66]).

The 'hierarchical approach' thus explains common linguistic expressions of the (un)truth predicate, which neglects the indication of orders, as being ambivalent. It tells us that we should not confuse the notion of (un)truth applicable to 2nd-order propositions with the notion of (un)truth applicable to 1st-order propositions. Distinguishing 'propositions' and notions related to them by their order is based on the reasonable requirement for their non-circular construction (see 2). Then, the 'proposition' P^B is false^2 (numerical superscripts indicate the order of the falsity 'predicate') because no false^1 1st-order 'proposition' believed by Bouleus exists.

A novel variant of the paradox was published by Brandenburger and Keisler [46] as the *Brandenburger-Keisler Paradox.* The sentence

AB: "Ann believes that Bob's assumption is wrong."

is unproblematically true or false, but truth/falsity cannot be consistently assigned to it if AB is embedded in the scope of "Ann believes that Bob assumes that":

P^{AB}: "Ann believes that Bob assumes that Ann believes that Bob's assumption is wrong."

The paradoxical self-reference is simply eliminated by typing:

$$Bel^4_a(As^3_b(Bel^2_a(Wr^1[\imath p^1.As^1_b p^1])))$$

where p^1 is as above; "a" and "b" denote Ann and Bob, respectively; As^k stands for "assume"; Wr^k stands for "wrong" (i.e. 'not true^k'), and $\imath$ is a 'descriptive operator' (a singularisation function). Thus, P^{AB} is of order 5 and so it does not occur in the range of p^1. The truth/falsity of P^{AB} is unproblematic, since it depends on Ann's belief^4 concerning Bob's assumption^3 concerning Ann's belief^2.[3]

I offer further discussion of the 'hierarchical approach' in 7.3 below. In the rest of this section, I use (a variant of) the well-known *Preface Paradox* (see Makinson [242]) to show that a mechanical application of type restrictions need not yet bring the final solution to a paradox.

According to the Preface Paradox, (i) the author x of the book b believes that all her statements in b are true and mistake free, for x and b's reviewers assiduously checked b; despite that, (ii) x writes an apology in b's preface for all the mistakes possibly remaining in b, because x knows that even the best checking does not guarantee that b is mistake free.

The paradox stems from two conflicting statements that are schematically and preliminary formalisable as follows (where p is a variable for 'propositions'):

(i) $Bel_x(\forall p(p \in b \rightarrow T(p)))$

(ii) $Bel_x(\exists p(p \in b \wedge \neg T(p)))$

[3] The reading according to which P^{AB} is not true or false but without a truth value also exists in my approach, e.g. $Bel^1_a(As^1_b Bel^1_a(Wr^1[\imath p^1.As^1_b p^1]))$.

According to the widely-accepted assumption of EL, no agent can believe two contradictory propositions (the Axiom of Consistence), so (i) and (ii) are incompatible.

Unlike the original setting of the Preface Paradox I will now assume that b may involve (i) or (ii). But I will not utilise the assumption that the Axiom of Consistence is not valid. I will rather point out that if 'propositions' are stratified in accordance with VCP, statements (i) and (ii) need not be in conflict. It is natural, even, to think that the statement from the preface is a 'meta-statement' speaking about statements among which it does not belong.

While statement (i) is x's 'meta-statement' that concerns b's statements, (ii) is a similar 'meta-statement' normally created during x's reflection of the whole of b's text. The trouble lies in that one cannot be sure whether the statement (ii) should be read as a $(k+2)$nd-order 'meta-statement' speaking about kth-order 'propositions' (expressed by b's sentences) and also the $(k+1)$st-order statement (i), or as a $(k+1)$st-order 'meta-statement' exclusively speaking about b's kth-order 'propositions'. Only the second reading is incompatible with EL's Axiom of Consistence. And casting doubts on the axiom is not a matter of typing.

7.3 Semantic paradoxes

To facilitate deeper discussion of epistemic paradoxes, I will recap and partly extend the results of my earlier writings on semantic paradoxes as solved within TTT.[4]

I distinguish two TTT approaches, while approach (A) makes the linguistic dependence on semantic notions explicit (and so the solution appears to be Tarskian), approach (B) relies on the *Principle of Non-contradiction* (*PNC*); here I follow Tichý [422].

(A)-approach

Approach (A) disambiguates the *'liar' sentence* (or another expression containing a semantic predicate)

S_L: "Sentence S_L is not true."

as representing a sentence such as

S_{L^1}: "Sentence S_{L^1} is not true in (the language) L^1."

[4] See e.g. Raclavský [333, 337, 323, 328], Raclavský, Kuchyňka and Pezlar [339].

The sentence seems to comment on the truth of a certain sentence in the language L^1, which presupposes that the sentence has a certain meaning in L^1. However, it can be realised that if L^1 is sufficiently well-specified, it does not allow the string "Sentence S_{L^1} is not true." to have the meaning it seems to have. As a result, the *Liar Paradox* is solved. (The solution to other semantic paradoxes is similar.)

Following Tichý [422], I will understand *language* here as a 'code':

Convention 35 (Language as a kth-order code, L^k)

By "*language* L^k" (as a 'code') I will call a partial function from numbers (Gödel numbers of expressions) to constructions of order k, i.e. as an object of type $(*_k\rho)$.

Remarks. In Raclavský [318], I show in detail how to define *language* in both *synchronic* and *diachronic senses.* Of course, one might object that the models are not the optimal models of natural language: language as such is not (a) a function from expressions to meanings but rather (b) a normative system enabling its users to communicate. But as already argued by Lewis [236], both (a) and (b) are possible treatments of language, while each of them models a certain important aspect of a language as such.

The Functional-constructional VCP (2.2.5) does not allow an expression "L^1" to have the 2nd-order construction ${}^{(0)}L^1$ as its meaning in L^1, where L^1 is of type $(*_1\rho)$. ${}^{(0)}L^1$, or any other construction v-constructing L^1, can be communicated, and so expressed by some of its expression, in a 2nd-(or a higher-)order language L^2, which involves (a translation of) L^1. L^2 is essentially richer than L^1 since expressions that could not bear meaning in L^1 can be meaningful in L^2. Similarly, there is L^3, L^4, etc., i.e. a whole *hierarchy of (meta)languages* forming one 'family', i.e. an object of type $(o(*_k\rho))$. (See also further conditions imposed on the construction of such a 'family' e.g. in Raclavský [333].)

No member of a 'family' allows a discussion of its own semantic properties: there are essential restrictions on what any language enables to communicate. Contrary to what is assumed in many philosophical writings, no semantic relation exists that links expressions with their meanings or denotata *simpliciter*. *Semantic relations*, e.g. TO BE A MEANING C^k OF AN EXPRESSION E IN L^k or TO BE A DENOTATUM D^τ OF AN EXPRESSION E IN L^k are relations(-in-extension) containing a language L^k. (Semantic relations conceived as contingently changing across time and logical space are conditions.)

For example, the relations of meaning and denotation are definable as follows (where e/ρ – the variable for Gödel numbers of expressions; $c^k/*_k; d^\tau/\tau;{}^{(0)}\mathbf{L}^k/\ (*_k\rho);\ {}^0{=}^{*_k}\ /(o*_k*_k);\ {}^0{=}^\tau\ /(o\tau\tau);{}^{(0)}\mathbf{Mean}^k/(o\rho*_k(*_k\rho))$; ${}^{(0)}\mathbf{Den}^k/(o\rho\tau(*_k\rho))$ – particular relations of denotation differ in particular τ):

Definition 95 (The meaning in L^k; the denotatum in L^k)

$$\models [\mathbf{Mean}^k\, e\, c^k\, \mathbf{L}^k] \Leftrightarrow_o [c^k =^{*_k} [\mathbf{L}^k\, e]]$$

$$\models [\mathbf{Den}^k\, e\, d^\tau\, \mathbf{L}^k] \Leftrightarrow_o [d^\tau =^\tau {}^2[\mathbf{L}^k\, e]]$$

Both ${}^0\mathbf{Mean}^k$ and ${}^0\mathbf{Den}^k$ can be communicated in a language of a (strictly) greater order than k, not in L^k or a language of even a lower order.

Remark. Sometimes a language in question is suppressed. In such a case we may read the expression "mean(ing)" as having an 'inbuilt' language, e.g. $\models [\mathbf{Mean}^{L^k}\, e\, c^k] \Leftrightarrow_o [\mathbf{Mean}^k\, e\, c^k\, \mathbf{L}^k]$, where ${}^{(0)}\mathbf{Mean}^{L^k}/(o\rho*_k)$. After such a disambiguation, one must sometimes employ the (B) method for solving paradoxes, see 7.3 below.

Comments.

i. Tarski [398] first expressed principal limitations on the possibility to communicate meanings, denotation and other *semantic relations of expressions* in a language in his famous *Tarski's undefinability theorem.*[5] Evidently, I accept the principle. I only add the idea that (natural) language can be modelled by a function from expressions to constructions of a certain order (or by a 'family' of such 'codes'), which gives Tarski's theorem backing in VCP. The 'hierarchical approach' to semantic paradoxes can then only be criticised with difficulties.

ii. In philosophical circles (e.g. Fitch [126], Kripke [220], Priest [306], Grim [160], Anderson [7], Tucker [431]), a certain nostalgia is often expressed: why cannot we have a 'global' ('universal') notion of truth that would decide all sentences of our vastly expressive language? Yet Tarski's theorem unequivocally says that we cannot.

Tarski's theorem does not imply that our language is not sufficiently 'global' and thus incapable of formulating statements such as "Every (total) 'proposition' is either true or untrue.". Such statements can, of course, be formulated within a sufficiently rich 'metalanguage-level' of our language that is modelled as a 'family' of codes. But Tarski's conception adds that a stronger 'metalanguage-level' of our language exists that is capable of speaking about the semantic properties of the former 'metalanguage-level', and that the above statement does not apply to the latter 'metalanguage-level'. I add that when one disputes a certain hierarchy H (say of languages), she speculates within another (meta)hierarchy MH enabling discussion of H.[6]

iii. Type restrictions are applied to 'propositions', since they have been identified with constructions. On the other hand, restrictions do not apply to propositions. For example, the following 1st-order construction contains the variable p that ranges over the type consisting of all propositions without

---

[5] The theorem was thoroughly investigated by Hilbert and Bernays [173], see also Priest [309].

[6] Details appeared in Raclavský, Kuchyňka and Pezlar [339].

exception (where $p/\pi; t/\rho; w/\omega;^{(0)} \neg/(oo);^{(0)} \vee/(ooo);^{(0)} \mathbf{Tot}^{\pi}/(\pi\pi)$ – the operator Tot (4.4.3) plays the role of the strong 'truth predicate' for propositions):

Definition 96 (Law of Excluded Middle for Propositions)

$\models \emptyset \Rightarrow [[\mathbf{Tot}^{\pi}\, p]_{tw} \vee [\neg[\mathbf{Tot}^{\pi}\, p]_{tw}]]$

Similarly, the type ranged over by e contains all sentences (their Gödel numbers). The *Law of Excluded Middle for Sentences*, however, has particular versions due to the fact that the truth of sentences depends on the truth of constructions of various orders the sentences express in this or that language (cf. the definition below); this brings restrictions back.

iv. *Semantic relations* must not be confused with *syntactic relations.* While semantic relations are typically objects of type $(o\rho(*_k))$ (relations-in-extension between expressions and languages-as-codes) or $(o\rho(o(*_k)))$ (relations-in-extension between expressions and 'families' of languages-as-codes), syntactical relations are typically objects of type $(o\rho(o\rho))$ (relations-in-extension between expressions and sets of expressions).

For example, the relation of provability between formulas and axiomatic theories (understood as sets of expressions) is not restricted (of course, until one imposes certain restrictions because of certain semantical properties of the expressions). It is thus possible to formulate an instance of the *Diagonal lemma,* e.g. $\varphi \leftrightarrow \neg Pr(\overline{\varphi})$, where "$\overline{\varphi}$" is the Gödel number of the formula "φ" and "Pr" is the predicate of provability in a given axiomatic theory. Note, however, that there is also a diagonal lemma concerning constructions – the existence of such a lemma depends on the *GPR.*

v. Kripke [220] published an influential criticism of the 'hierarchical approach' to truth. He attacked the very idea of disambiguation by pointing to the fact that a. the official English language does not contain explicitly indexed predicates "true$_{L^1}$", "true$_{L^2}$", etc. (and so the 'hierarchisation' is an *ad hoc* method), and that b. the distribution of indices is questionable in cases such as "[Plato says:] Everything Socrates says is true. [Socrates says] Everything Plato says is false [or: not true]." (the *Buridan Paradox*) or "[speaker in room no. 231:] Everything I have said in room no. 231 is false." (the *Contingent Liar Paradox*).

With regards to case a., the 'hierarchical approach' can be defended by the reference to the fact that some English expressions, e.g. "not true in Czech", explicitly relate truth to language.[7] To illustrate, the following sentence is perfectly natural: " 'Pivo je horké.' is true in Slovak but not in Czech.", for it says that the cited sentence "Pivo je horké." is true in Slovak (for in Slovak it

[7] Which was mentioned as a lack of many contemporary approaches to truth by Gupta and Belnap [161]; cf. the recent theories of truth e.g. in Achourioti, Galinon, Martínez and Fujimoto [2].

means that beer is bitter), but it is not (normally) true in Czech (for in Czech it means that beer is hot). More importantly, when reconstructing our everyday linguistic communication, we must employ disambiguation and other 'artificial means' to eliminate contradictions, though our model will cease to bear a 1-1 correspondence with the modelled reality (which itself often appears vague and sometimes intuitively contradictory), it is a matter of adequate choices.

Remarks. Of course, one need not deploy any artificial tool if we disregard Tarski's assumption that the language into which we introduce a T-predicate is sufficiently rich (Raclavský [333]). A well-known example of such an expressively weak language is the language of three-valued logic $\mathcal{L}_{(TFN)}$ in which the weak falsity predicate is not allowed to be reworked into the strong falsity predicate (cf. 4.5.5). It is well known that in such an approach, the Liar Paradox is solved by considering both the liar sentence and its negation to have the third value, N, so the paradox is solved. If, however, the indicated increase of $\mathcal{L}_{(TFN)}$'s expressive power were allowed, the paradox would strike back. The approach thus suffers from the *revenge problem*, namely the existence of the *Strengthened Liar Paradox*.[8] Thus, Kripke [220] admitted that the 'ghost of Tarski's hierarchies' is still with us.

This is even confirmed for the putative *revenge Liar for 'hierarchical solution'*.[9] The natural candidate for the respective revenge-Liar-like sentence seems to be "This sentence is not true at any level." (*RHL*, for short). However, if the unclear notion of level is understood as an analogue of the notion of language and we explicate it as suggested above, RHL i. lacks meaning in any language in the scope of the variable for languages L^1 – L^n of which RHL is supposed to be a member and ii. it has a certain meaning, and is true, in a metalanguage ML^i (for $n < i$) that is capable to speak about the languages L^1 – L^n. Similar consideration applies to the variant of RHL, "This sentence is not true in any of metalanguages ML^i.".

(B)-approach

Now I am going to present the approach based on the PNC. It excludes the possibility that a proposition (or 'proposition', or sentence) is both true and false, yet it does not exclude being gappy – so PNC is partly classical, partly not. Here is a definition of the propositional variant of the law:

---

8 The revenge problem for a particular approach to a paradox consists in the existence of a *revenge form of the paradox* that discredits the key feature of the approach by showing the novel paradox is not solvable by the approach (see e.g. Beall [32]). For example, the Strenghtened Liar Paradox presents the *Revenge Liar Paradox* for the three-valued approach to the Liar Paradox.

9 The revenge paradox raised against the 'hierarchical solution' to FP can be well documented in literature; it is discussed in 7.7.3 below. For the case of the Liar paradox, both Burge's [53] contextualistic hierarchical solution, similarly as the situation semantics of Barwise and Perry [27], was vulnerable to the criticism that it cannot express certain 'global' notions (see e.g. Grim [159], [431]).

Definition 97 (Law of Non-contradiction for Propositions, PNC)

$\models \emptyset \Rightarrow [\neg[[\mathbf{Tot}^{\pi}\, p]_{tw} \wedge [\neg[\mathbf{Tot}^{\pi}\, p]_{tw}]]]$

Tarski [398] first used the classical PNC when proving Tarski's theorem, because he used the very possibility of the construction of the Liar Paradox within a given language L for demonstration that L in fact cannot communicate the notion of falsity-in-L. Tichý [422] deployed PNC when elaborating his suspicion that Russell's sole VCP does not solve the Liar Paradox.

To show it, I first define the strong notion of truth applicable to (o-)constructions (where $o/o; c^k/*_k;^{(0)}\,\mathtt{T}/o;^{(0)}\,\mathbf{Tr}^k$ (to be a true 1st-order construction) $/(\pi *_k);^{(0)} =^o /(ooo);\ ^{(0)}\exists^o/(o(oo)))$.

Definition 98 (Strong truth for kth-order constructions)

$\models [\mathbf{Tr}^k\, c^k]_{tw} \Leftrightarrow_o [\exists^o[\lambda o[[^2c^k =^o\ o] \wedge [o =^o \mathtt{T}]]]]$

Remarks. The strong 'truth predicate' for constructions somewhat resembles the (type-theoretically comparable) operator Tot (3.3.3). My definiens is v-congruent with the definiens proposed by Tichý [422]. The definiens does not include the variables t and w, which can, however, occur freely in the construction that is the value of c^1, and so they are free in $^2c^1$. The 'predicate' assigns $\mathtt{F}$ to all constructions (i.e. it decides them to be false) that do not v-construct the truth value $\mathtt{T}$, which includes v-improper o-constructions and also constructions of objects of types different from o (e.g., constructions of sets of numbers). A 'truth predicate' that would not decide the latter type of constructions as true or false (so they would not fall into extension or anti-extension of the 'predicate') is also definable (where $^{(0)}\mathbf{Tr}^{PTk}/(\pi *_k)$; adapted from Raclavský [323]):

Definition 99 (Weak truth for kth-order constructions)

$\models [\mathbf{Tr}^{PTk}\, c^k]_{tw} \Leftrightarrow_o [\mathbf{Tot}^{\pi}\, {}^2c^k]_{tw}$

I also add the rule corresponding to T-axiom (where $^0/C//(*_k *_k)$ – the alethic normalisation function, 6.4):

Example 76 (The Rule of Veridicality of T)

$$\frac{[\mathbf{Tr}^k\, c^k]_{tw} \cong o}{^2/c^k/ \cong o}$$

Here, I adapt Tichý's [422] definition of the notion of the 1st-order liar. Let $x/\iota;^{(0)}\,\exists^{*_1}/(o(o*_1));^{(0)}\,\mathbf{Li}^1$ (to be a 1st-order liar)$/(\pi\iota);^{(0)}\,\mathbf{As}^1$ (to assert a 1st-order construction) $/(\pi\iota *_1);^{(0)}\,\mathbf{Tr}^1/(\pi *_1)$.

Definition 100 (1st-order Liar)

$\models [\mathbf{Li}^1\, x]_{tw} \Leftrightarrow_o [\exists^{*_1} [\lambda c^1 [[\mathbf{As}^1\, x\, c^1]_{tw} \wedge [\neg[\mathbf{Tr}^1\, c^1]_{tw}]]]]$

Now Russell [353, 445] seems to consider the meaning of "x is a liar" to be the 'propositional function' that resembles my definiens above. Russell solved the Liar Paradox by referring to the fact that the 2nd-order 'proposition', which comes from insertion of an individual in place of the variable x contained in the 2nd-order 'propositional function', cannot be a value of a 'propositional' variable (in our case c^1) contained in it, because, according to VCP, the variable's range consists exclusively of 1st-order 'propositions'.

However, Tichý [422] noted that the lowest possible order of the above definiendum equals 1, unlike the definiens, whose lowest possible order equals 2. The 1st-order construction $[\mathbf{Li}^1\, \mathbf{X}]_{tw}$ (where ${}^{(0)}\mathbf{X}/\iota$) thus occurs in the range of the variable c^1, there is no apparent violation of VCP. This fact led Tichý [422] to the suspicion that Russell ([353], [445] with Whitehead) was wrong in considering VCP as a universal cure to logical and semantic paradoxes.

Remark. In defence of Russell, I maintain that the 1st-order construction $[\mathbf{Li}^1\, \mathbf{X}]_{tw}$ is (because of its definition) related towards the 2nd-order construction $[\exists^{*_1} [\lambda c^1 [[\mathbf{As}^1\, c^1]_{tw} \wedge [\neg[\mathbf{Tr}^1\, c^1]_{tw}]]]]$, which presupposes that ${}^{(0)}\mathbf{As}^1$ has a non-circular definition. Since the notion of assertion of a certain 1st-order construction C is not primitive, asserting C requires that both speaker and addressee of an assertion-event employ a certain communicational system, which can be simply understood as L^1. It thus follows from the considerations given above that VCP is also involved in this case.

Tichý's [422] solution to the respective version of the Liar Paradox rather employs PNC: no individual X can assert the construction $[\mathbf{Li}^1\, \mathbf{X}]_{tw}$, for if it could happen at some T and in some W, the construction saying that X asserts an untrue construction would be true at T and in W, which contradicts the assumption. Tichý's solution is in fact a compact variant of Prior's [310] (who discussed different versions of the Liar); see also 7.4 where I return to the issue.[10]

To sum up the main idea of the just discussed (B)-approach to semantic paradoxes, sentences that speak about semantic notions, but do not contain an explicit reference to a certain language, refer to a certain language implicitly. According to both approaches (A) and (B), it holds that, though the corresponding 'liar sentences' have certain meanings (they are not meaningless), they do not enable a construction of a genuine paradox. For example,

---

[10] In recent discussions, Prior's observations are studied intensively, see e.g. Priest [307] (who contended that it is more natural to give up PNC), Tucker and Thomason [432], Bacon, Hawthorne and Uzquiano [20], Bacon and Uzquiano [19], Tucker [431], Orilia and Landini [281].

the sentence "X is a liar." has a certain meaning in L^1, namely $[\mathbf{Li}^1\,\mathbf{X}]_{tw}$, and most users of L^1 can communicate the meaning without any obstacle. Except for X, who cannot normally use the sentence "X is liar." for expressing the intended meaning in L^1. This somewhat surprising claim is supported in the following two sections.

7.4 Limitations explication of 'propositional' notions

Davies [92] published a surprising paradox related to logical space ω, which is usually called *Kaplan's paradox* ("Kaplan's paradox about possible worlds"). The paradox has been studied by a number of scholars, e.g. Grim [159], Oksanen [279], Lindström [237], Anderson [7], and even Kaplan [203] who suggested the possibility to solve it by a hierarchy of models of levels $1 - n$.

The paradox is based on a consequence of Cantor theorem, according to which the cardinality of any set is strictly lower than the cardinality of its power set, e.g. $|\,\omega\,| < |\,\mathcal{P}(\omega)\,|$. (When speaking about ω, we in fact identify it with the universal set of ω-objects, i.e. an object of type $(o\omega)$.) By Cantor theorem, if $|\,\omega\,| = N$, the number of propositions $|\,\mathcal{P}(\omega)\,| = 2^N$. But, since one can have an attitude towards an arbitrary proposition, there is a corresponding amount of state-of-affairs involving propositions about such attitudes, and so there are 2^N possible worlds – which, however, contradicts the assumption that there are N possible worlds.[11]

Following e.g. Tichý [422], I understand $|\,\omega\,| < |\,\mathcal{P}(\omega)\,|$ rather as producing an inevitable limitation of logical space:

Theorem 2 (Limitation of logical space I)

Every logical space ω is limited, for there is no 1-1 correspondence between its members and its subsets or propositions.

[11] Grim [159] attempted to use the paradox for supporting the claim that there is no set of all truths. He argued that to each member of the power set $\mathcal{P}(\Pi)$ of the set Π of truths-as-propositions $P_1, ..., P_n$ there corresponds a 'new' truth saying that P_i belongs to a certain member of Π. However, Grim overlooked that the proposition we denote by (say) "$P_1 \in \{P_1, P_2\}$" is simply one of $P_1, ..., P_n$, it is not a new proposition; consequently, there is no 'explosion' of truths, as he imagined. (On the other hand, if we take into account 'structured propositions' of RTT, there is no 'totality' of such truths, as already mantained by Russell [353].)

I utilise the theorem to start a wider investigation of representatives of an important group of modal notions applicable to 'propositions' and so related to propositions, which I call *'propositional' notions.*[12]

Convention 36 ('Propositional' notion)

By "*'propositional' notions*" I refer to notions such as knowledge, belief, truth, assertion, etc., which v-construct monadic, dyadic, etc., conditions that are satisfiable by individuals (or their n-tuples) and o-constructions (or their m-tuples).

Remark. Recall that medadic conditions of type $(\pi *_k)$ are determined by the 'propositional' notions such as ${}^{(0)}\mathbf{Tr}^k$ or ${}^{(0)}\Box^k$, while dyadic conditions of type $(\pi\iota *_k)$ are determined by ${}^{(0)}\mathbf{Bel}^k, {}^{(0)}\mathbf{K}^k$ or ${}^{(0)}\mathbf{As}^k$.

The aforementioned consequence of Cantor theorem is related to a limitation that has a great impact on our topic, since belief attitudes relate individuals to propositions:

Theorem 3 (Limitation of logical space II)

For every logical space ω (or: universal set of $\langle T, W\rangle$ couples) and at least a two-membered domain of individuals ι it holds that there is no 1-1 correspondence between propositions and m-ary conditions, for $1 < m$.

Remark. Proof hint: even if we considered sets ω and ι, each of cardinality 2, and admitted only total propositions (as objects of type $\pi := (o\omega)$), there would only be 4 propositions while there would be many monadic conditions of type $(\pi\iota)$ (namely 16) and there would even be more m-ary conditions ('relations') of type $(\pi\iota\pi)$.

The rationale behind the theorem can be explained as follows. Since propositions a. 'describe' parts of worlds in which they are true, and b. they are not independent of each other, not every combination of propositions 'describes' a part of a possible world. An exemplary pair of such incompatible propositions was mentioned in the final part of the previous section 7.3:

a. XENIA IS A LIAR

b. XENIA ASSERTS THAT SHE IS A LIAR, THE 'PROPOSITION' WHICH IS NOT TRUE.

[12] An investigations of some modal notions (applicable to names of sentences) within one framework already occurred in Pailos and Rosenblatt [286] (their investigation was anticipated by Halbach [162] and Paseau [291]); they adopt a different framework in which typing of predicates is not based on type theory (although it resembles it), but they nevertheless show that such typing solves *modal paradoxes* such as the Knower Paradox.

In formulating the following (meta-)theorem, I make use of the fact that for any o-construction C there exists just one *corresponding proposition* P (as I will call it) which has, in any particular T and W, always the same truth value (if any) which is $v(T/t; W/w)$-constructed by C (where $p/\pi;^{(0)} =^{\pi} /(o\pi\pi);^{(0)}$ $\mathbf{Corresp}^k/(o\pi *^k)$):

Definition 101 (Proposition corresponding to a construction)

$\models [\mathbf{Corresp}^k\, p\, c^k] \Leftrightarrow_o [p =^{\pi} [\lambda t \lambda w.{}^2c^k]]$

Theorem 4 (Limitation of logical space III)

The limitation of logical space mentioned in the theorems of type I and II is replicated on the level of o-constructions thanks to the link between constructions and propositions corresponding to them.

As a proof, I offer the following (it is adaptated from Tichý [422]). Let a particular moment of time T be fixed, and so when speaking about truth of constructions, or their v-constructing, I will not mention it at all. For further simplification, consider only two possible worlds, W_1 and W_2. Now every o-construction is of one of 9 'kinds' E_1 – E_9, depending on what it v-constructs:

	E_1	E_2	E_3	E_4	E_5	E_6	E_7	E_8	E_9
W_1	T	T	T	F	F	F	_	_	_
W_2	T	F	_	T	F	_	T	F	_

Remark. For example, all o-constructions that v-construct T in both W_1 and W_2 are of 'kind' E_1, whereas all o-constructions v-constructing T in W_1, but F in W_2, are of 'kind' E_2, etc. The classification of constructions into 'kinds' E_1 – E_9 copies the classification of propositions into such 'kinds'.

Further, consider two o-constructions C^1 and D, while D speaks about a certain attitude R of an agent X towards C^1 which is not true:

a. $C^1 := [\mathbf{Lh}^1\, \mathbf{X}]_{tw}$

b. $D := [[\mathbf{As}^1\, \mathbf{X}\, {}^0C^1]_{tw} \wedge [\neg[\mathbf{Tr}^1\, {}^0C^1]_{tw}]]$

The following table demonstrates that D is never v-congruent with C^1.

If W is ...	*and X is related by R to C^1 of kind ...*	*then D is of kind ...*
W_1	E_1, E_2, or E_3	E_4, E_5, or E_6
	E_4, E_5, E_6, E_7, E_8, or E_9	E_1, E_2, or E_3
W_2	E_1, E_4, or E_7	E_2, E_5, or E_8
	E_2, E_3, E_5, E_6, E_8, or E_9	E_1, E_4, or E_7

Remark. The same holds for $D' := [\exists^{*1}\lambda c^1[[\mathbf{As}^1\,\mathbf{X}\,c^1]_{tw} \wedge [\neg[\mathbf{Tr}^1\,c^1]_{tw}]]]$, i.e. the definiens of $[\mathbf{Li}^1\,\mathbf{X}]_{tw}$, but in that case, the third column does not involve E_6, E_3, E_8, E_7.

That C^1 cannot be a construction an agent X is related to by the assertion relation (as I mentioned at the end of 7.3) is thus a consequence of the limitation of any possible explication of 'propositional' notions, which follows from the limitation of logical space. By generalising my consideration, in no possible world can proposition P, corresponding to C^1, overlap with proposition Q that corresponds to D: a possible world in which P and Q (and thus also C^1 and D) would have the same truth value clearly does not exist.

The same findings pertain to different, yet similar pairs of constructions. For example (where ${}^0\mathbf{Fo}^1$ (to be a fool) $/(\pi\iota)$):

a′. $C^1 := [\mathbf{Fo}^1\,x]_{tw}$

b′. $D := [[\mathbf{Bel}^1\,x\,{}^0C^1]_{tw} \wedge [\neg[\mathbf{Tr}^1\,{}^0C^1]_{tw}]]$

On the other hand, my findings do not apply to every 'propositional' notion. For example, the notion of knowledge is governed by the Rule of Factivity of K (6.4), in consequence of which the construction of form D, namely $[[\mathbf{K}^1\,x\,{}^0C^1]_{tw} \wedge [\neg[\mathbf{Tr}^1\,{}^0C^1]_{tw}]]$, is internally inconsistent, as can easily be checked. The notion of knowledge is, nevertheless, subjected to another interesting type of limitation, which I am going to discuss in the next (sub)section.

Thus, the above considerations can be understood as a proof of the following theorem (for the case of assertion specifically, see Tichý [422]):

Theorem 5 (Limitation of explication of 'propositional' notions such as belief and assertion)

Some 'propositional' notions, e.g. *As* or *Bel*, which are not governed by the T-axiom, cannot be explicated so that an agent X could bear the assertion/belief relation towards an o-construction C such that i. C is v-congruent with construction D and ii. D says that X asserts/believes C, while C is not true.

Remark. Wiśniewski [451] proved that even mild assumptions lead to the conclusion that they are recursive propositions that cannot by assigned to any sentence of a language. For a study of inexpressible relations and claims, see also Fritz [144]. Within unramified multimodal logic, Anderson [7] proved inexpressibility of the Epimenidean liar sentence for Epimenides. On the other hand, Thomason [409], Priest [307] and Kaplan [203] mentioned a possibility of inexpressible 'propositions' with a dismissal.

It can be easily noted that paradoxes such as the Bouleus Paradox (7.2) or the Liar Paradox (7.3) are caused by our mistaken understanding of their crucial sentence S as expressing a construction of kind D, where D says that an agent has an attitude towards an untrue construction C that is v-congruent with D. But this is only a confusion of our logical intuitions, since, strictly speaking, no X can express such a construction D using S.

Remarks. Asher [16] raised a worry about free logic that suggests (as I do, given the application of the theorem) that the Cretan uttering "Everything I say is false." (henceforth S^C) does not manage to express a 'proposition' in the circumstance in which the token of S^C is his sole utterance. Thus, the unexpressed 'proposition' cannot be the target of anaphoric reference, and so Socrates' response to Cretan's utterance by uttering "I don't believe that." does not seem to posses any belief – contrary to our intuition. Note, however, that Cretan's utterance at least indirectly served him to point at the intended meaning (a 'proposition'), and that meaning presents the content of Socrates' belief.

Clearly, S^C has a certain meaning in English: the 'proposition' exists and it is unaffacted by any contingent fact. Semantic representations of sentences are thus graspable and can be calculated by any speaker competent to English, as required by Thomason [409]. The only limitation we face here is the impossibility for the Cretan to bring about, by uttering S^C, a world in which the only 'proposition' he is related to by the assertion relation is the one expressed by S^C. The theorems about the limitation of logical space II and III tell us that that is impossible.[13]

7.5 The Paradox of Impossibility of Knowing One's Own Ignorance

Recall that by the General Principle of Reducibility (2.4.4), the type of (say) 1st-order constructions is richer than it might appear given the restrictions enforced by the non-circular specification of construction (VCP). Questions may then arise as to whether the present approach is immune to paradoxes, and how it relates to the considerations of the preceding section.

I expose a paradox resembling the Knower Paradox (see 7.6 below), which reveals the fact that notions such as knowledge are affected by the limitation of logical space.

[13] Prior [310] was right when referring to the world as 'odd' for it does not fit the paradoxical scenario. I add that the brute logical facts (see the above theorems) give us no hope with regards to the possibility to change the 'oddity'.

Theorem 6 (Limitation of explication of 'propositional' notions such as knowledge)

Some 'propositional' notions, e.g. knowledge, which are governed by the T-axiom, cannot be explicated so that an agent X could bear the relation of knowing towards an o-construction C such that i. C is v-congruent with the construction D and ii. D says that X does not know C.

As a proof, I exhibit a 1st-order o-construction which I will denote by "$\boxed{K^1}$" (where ${}^{(0)}\boxed{K^1}/o$). This is nevertheless only a handy abbreviation for the construction that says that an individual which is the value of x has a certain nameless property G. Similar convention will be assumed for the case of $\boxed{K^1}$'s mates such as $\boxed{Bel^1}$ and $\boxed{As^1}$.

Convention 37 (Construction $\boxed{K^1}$)

"$\boxed{K^1}$" is short for

"$[G\,x]_{tw}$"

where $x/\iota; t/\rho; w/\omega;{}^{(0)}\, G/(\pi\iota)$.

$\boxed{K^1}$ is v-congruent with a 2nd-order construction of which $\boxed{K^1}$ is a 1st-order subconstruction and that says that $\boxed{K^1}$ is not known.

Definition 102 (Construction $\boxed{K^1}$)

$\models \boxed{K^1} \Leftrightarrow_o [\neg[\mathbf{K}^1\ x\ {}^0\boxed{K^1}]_{tw}]$

Remark. This definition is obviously, but not viciously circular (for a theory of circular definitions, see Gupta and Belnap [161]). For those who do not like the idea of circular definitions, the 'definition' and its few mates below are simply bi-directional rules that are derivable from the Diagonal lemma that concerns constructions.

It is easy to get a theorem according to which the o-construction $\boxed{K^1}$ cannot be known[1] (i.e. known in the sense of $\mathbf{K}^1$). This can be easily adjusted for other orders. Let ${}^{(0)}\forall^\iota/(o(o\iota));{}^{(0)}\underset{o}{\Box}/(o\pi)$ (see 6.3.1):

Theorem 7 (Impossibility of knowing the o-construction $\boxed{K^1}$)

$\vdash \emptyset \Rightarrow [\forall^\iota \lambda x[\underset{o}{\Box}[\lambda t\lambda w[\neg[\mathbf{K}^1\ x\ {}^0\boxed{K^1}]_{tw}]]]] \cong \mathtt{T}$

For the sake of the *reductio* proof, assume that $\boxed{K^1}$ v-constructs the truth value $\mathtt{F}$. Thanks to definition of $\boxed{K^1}$, $[\neg[\mathbf{K}^1\ x\ {}^0\boxed{K^1}]_{tw}]$ v-constructs $\mathtt{F}$, too.

On the basis of $\neg$, $[\mathbf{K}^1\ x\ {}^0\boxed{K^1}]_{tw}$ v-constructs T.[14] In accordance with the (intuitive) Rule of Factivity of K we derive from it that $\boxed{K^1}$ v-constructs T, which, however, contradicts the assumption that $\boxed{K^1}$ v-constructs F. Thus, $\boxed{K^1}$ v-constructs the opposite truth value, namely T, and so, thanks to definition of $\boxed{K^1}$, $[\neg[\mathbf{K}^1\ x\ {}^0\boxed{K^1}]_{tw}]$ v-constructs T, too.

The theorem entails that the construction $\boxed{K^1}$ (and thus also its definitional equivalent) is a valid construction.

Corollary 7.1

i. $\vdash \emptyset \Rightarrow \boxed{K^1} \cong \mathtt{T}$

ii. $\vdash \emptyset \Rightarrow [\neg[\mathbf{K}^1\ x\ {}^0\boxed{K^1}]_{tw}] \cong \mathtt{T}$

Remark. Literature presents us with distinct conceptions of verificationism (e.g. Martin-Löf [247], see also 7.7.4) according to which every truth is knowable, but a plain verificationism seems to be untenable given the corollary.

The fact that no agent can know a construction such as ${}^{(0)}\boxed{K^1}$ sounds paradoxical; I will call the corresponding paradox the "*Paradox of Impossibility of Knowing One's Own Ignorance*".

Now there is an important lemma:

Lemma 6.1 (Invalidity of the naive Rule of Factivity of K)

The existence of constructions such as $\boxed{K^1}$ causes the unrestricted Rule of Factivity of K to be unsound.

For a proof consider (a) $[\mathbf{K}^1\ x\ {}^0\boxed{K^1}]_{tw} \cong \mathtt{T}$ which the naive Rule of Factivity of K simplifies to (b) ${}^0\boxed{K^1} \cong \mathtt{T}$. Since (b) is a validity (by Corollary 6.1), (a) must be false (by definition of the construction $\boxed{K^1}$).

The naive (unrestricted) rule thus has to be restricted, as I maintained in 6.4: reasoning with e.g. $\boxed{K^1}$ while using the Rule of Factivity of K is not permitted. Note that there are infinitely many constructions involving $\boxed{K^1}$:

a. constructions with 'extensionally embedded' $\boxed{K^1}$, e.g. $[\boxed{K^1} \vee C^1]$

b. constructions with 'intensionally embedded' $\boxed{K^1}$, e.g. $[\mathbf{Bel}^1\ x\ {}^0\boxed{K^1}]_w$

[14] Here I assume that if $\boxed{K^1}$ is an o-construction, $[\mathbf{K}^1\ x\ {}^0\boxed{K^1}]$ v-constructs a proposition that is not partial (this is a natural assumption with regards to K), and so, for any T and W, its value is either T, or F.

while only case a., and case b. only with specific examples such as e.g. $[\mathbf{K}^1\, x\, {}^0\boxed{K^1}]_{tw}$, negatively affects the Rule of Factivity of K. It is then tempting to maintain that one cannot substitute in the rule of factivity of K a construction which contains $\boxed{K^1}$ as a direct subconstruction. However, there are equally dangerous constructions, in which $\boxed{K^1}$ is not a direct subconstruction, cf. e.g. ${}^2[\mathrm{Id}^k\, {}^0\boxed{K^1}]$ (where ${}^{(0)}\mathrm{Id}^k$ (identity function) $/(*_k *_k)$ – see 2.5.1), in which v-constructing $\boxed{K^1}$ is executed. This is why I speak about limitation of the factivity rule of K so that a construction executing $\boxed{K^1}$ during its v-constructing cannot be substituted.

Remark. Note that the Rule of Factivity of K is only negatively affected by certain constructions whose definitions require certain higher-order constructions, while the GPR is employed. This fact relates to Myhill's [272] and Giaretta's [154] elaboration of the suspicion that RTT is incapable of avoiding Grelling's Paradox/the Liar Paradox if the Axiom of Reducibility is adopted, for in such a case, the initial formal language is extended so that the paradox can be reconstructed in it. For illustration, recall the construction ${}^{(0)}\mathbf{Li}^1$, whose name extends the initial formal language of a hierarchy of languages.

When using an adaptation of the above $\boxed{K^1}$-related proof, it can be easily found that an analogous problem – hence briefly referred to as "$\boxed{P}$-*problem*" – also affects other 'propositional' notions, e.g. belief and assertion (where ${}^{(0)}\boxed{Bel^1}, {}^{(0)}\boxed{As^1}/o$).

Definition 103 (Constructions $\boxed{Bel^1}$ and $\boxed{As^1}$)

$$\models \boxed{Bel^1} \Leftrightarrow_o [\neg[\mathbf{Bel}^1\, x\, {}^0\boxed{Bel^1}]_{tw}]$$

$$\models \boxed{As^1} \Leftrightarrow_o [\neg[\mathbf{As}^1\, x\, {}^0\boxed{As^1}]_{tw}]$$

Theorem 8 (Impossibility to i. believe the o-construction $\boxed{Bel^1}$ and ii. assert the o-construction $\boxed{As^1}$)

i.
$$\vdash \emptyset \Rightarrow [\forall^\iota \lambda x[\underset{o}{\Box}[\lambda t \lambda w[\neg[\mathbf{Bel}^1\, x\, {}^0\boxed{Bel^1}]_{tw}]]]] \cong \mathrm{T}$$
ii.
$$\vdash \emptyset \Rightarrow [\forall^\iota \lambda x[\underset{o}{\Box}[\lambda t \lambda w[\neg[\mathbf{As}^1\, x\, {}^0\boxed{As^1}]_{tw}]]]] \cong \mathrm{T}$$

The theorems have quite analogous corollaries that contain definitional equivalents of the constructions $\boxed{Bel^1}$ and $\boxed{As^1}$.

Remarks. Part ii. of the theorem can be understood as an extreme case of *Moore's paradox* (Moore [264]), according to which an agent expresses the belief that it is raining, but he does not believe that it is raining. This invites the question concerning coherence of his thinking, as argued by Hintikka [177] and expressed in EL by the well-known Axiom of Consistence. The impossibility to assert $\boxed{As^1}$ or $[\neg[\mathbf{As}^1\ x\ {}^0\boxed{As^1}]_{tw}]$ also resembles a paradox. Consider a sentence such as "I do not assert anything by this sentence.", or "This sentence has no meaning." of the forgotten *Paradox of the Non-communicator* by Drange [98].

On the other hand, such claims do not apply to the notions of necessity and truth. To see it, let us first define the notion ${}^{(0)}\square^k$ that is applicable to kth-order constructions (where ${}^{(0)}\square^k/(o*_k)$):

Definition 104 (Necessity of o-constructions, $\square^k$)

$$\models [\square^k\, c^k] \Leftrightarrow_o [\underset{o}{\square}\,[\lambda t\lambda w.{}^2c^k]]$$

It can be easily checked that straightforward $\boxed{P}$-forms of constructions $[\neg[\square^1\, c^k]]$ and $[\neg[\mathbf{Tr}^1\, c^k]]$ are not definable. If $\boxed{Tr^1}$ v-constructed $\mathtt{T}$, then its alleged definiens $[\neg[\mathbf{Tr}^1\boxed{Tr^1}]]$ would v-construct $\mathtt{F}$, and *vice versa*. Similarly for the case with $\square^k$.

Theorem 9 (Undefiniability of $\boxed{P}$-forms of constructions expressing non-necessity/untruth)

$\boxed{P}$-forms of constructions expressing i. non-necessity, ii. untruth are not definable.

We must naturally ask why the notion of knowledge behaves differently from the notions of necessity or truth, although they are all governed by the same T-axiom.

My explanation traces their difference to the fact that the set of propositions cannot be mapped in 1-1 fashion to the members of logical space ω and the related theorems (7.4):

1. The notions of necessity or truth apply to propositions, or to constructions (to which certain propositions correspond). The two 'propositional' notions are thus only affected by the cardinality of logical space ω, since the cardinality of ω has merely an impact on the nature of propositions (how many worlds they contain, how many propositions exist, etc.). Thus, the notions of necessity and truth are rather unaffected.

2. On the other hand, notions that deal with conditions satisfiable by individuals and propositions, or by individuals and constructions with which they correspond certain propositions, are markedly affected by limitation of logical space, cf. theorems about the limitation of logical space II and III.

The fact that not every imaginable linkage between a proposition and a condition is indeed logically possible is made especially visible by the Liar Paradox and the Paradox of Impossibility of Knowing One's Own Ignorance: it is not possible to match e.g. the proposition that corresponds to the construction $\boxed{K^1}$ with the proposition that corresponds to the construction $[\mathbf{K}^1\, x\, {}^0\boxed{K^1}]_{tw}$. Inevitably, the limitation passes to the level of constructions, cf. Theorem about the limitation of logical space III.

Here is a table summarising our key findings concerning 'propositional' notions.

A 'propositional' notion C^P	T*-axiom*	$\boxed{P}$*-problem*	C^P v*-constructs an object of type*
${}^{(0)}\mathbf{Tr}^k$	✓	×	$(\pi *_k)$
${}^{(0)}\Box^k$	✓	×	$(\pi *_k)$
${}^{(0)}\mathbf{K}^k$	✓	✓	$(\pi \iota *_k)$
${}^{(0)}\mathbf{Bel}^k$	×	✓	$(\pi \iota *_k)$
${}^{(0)}\mathbf{As}^k$	×	✓	$(\pi \iota *_k)$

I sum up by accentuating the above observation that not every thinkable combination of 'propositions' can characterise a realisable possible world:

3. Some of our intuitive possible worlds (e.g. the world where one truly asserts that she is a Liar) are only determination systems (4.3.1), not possible worlds in the proper sense of the word.

7.6 The Knower Paradox

The above Paradox of Impossibility of Knowing One's Own Ignorance resembles the *Knower Paradox* ("The Paradox of the Knower") that was published by Kaplan and Montague [206] and formally elaborated by Montague [261].[15] The two paradoxes are only partly similar, but not identical, especially since Montague's paradox utilises a syntactic predicate of knowledge "K", not the knowledge operator $\mathbf{K}^k$ applicable to o-constructions, and a different instance of the Diagonal lemma is used.

[15] Thomason [406] extended Montague's and Kaplan's results for the case of (rational) belief; for further investigation see e.g. Koons [218], Turner [434], Tymoczko [436]. For an investigation of self-reference in modal logic (and Löb theorem), see e.g. Smoryński [377].

The Knower Paradox utilises minimum rules governing "K" and can be condensed to the following, Liar-like sentence:

"Nobody knows this sentence.".

The following exposition of the paradox is adopted from Égré [114].

Example 77 (The Knower Paradox)

Let us extend Robinson's arithmetic Q to Q′ by extending, first, its language by Gödel numbers $\overline{\varphi}, \overline{\psi}$, etc., of formulas φ, ψ, etc., and also the following predicates:

"K(n)" which reads 'the truth of the formula with the Gödel number n is known',

"I(n, m)" which reads 'the formula with the Gödel number m is derivable from the formula with the Gödel number n' (this relation is definable in Q′).

Q′ extends Q by the following three axioms:

(T) $\mathrm{K}(\overline{\varphi}) \rightarrow \varphi$

(U) $\mathrm{K}(\overline{\mathrm{K}(\overline{\varphi}) \rightarrow \varphi})$

(I) $(\mathrm{I}(\overline{\varphi}, \overline{\psi}) \wedge (\mathrm{K}(\overline{\varphi})) \rightarrow \mathrm{K}(\overline{\psi})$

The Diagonal lemma is provable for Q′.

(D) $\vdash \varphi \leftrightarrow \mathrm{K}(\overline{\neg\varphi})$

Now we derive a contradiction from Q′:

1.	$\vdash \phi \leftrightarrow \mathrm{K}(\overline{\neg\phi})$	from (D) by substitution
2.	$\vdash \mathrm{K}(\overline{\neg\phi}) \rightarrow \neg\phi$	from (T) by substitution
3.	$\vdash \phi \rightarrow \neg\phi$	from 2 by employing 1
4.	$\vdash \neg\phi$	by 3
5.	$\vdash \mathrm{I}(\overline{2}, \overline{\neg\phi})$	from what precedes
6.	$\vdash \mathrm{K}(\overline{2}))$	by (U)
7.	$\vdash (\mathrm{K}(\overline{2}) \wedge \mathrm{I}(\overline{2}, \overline{\neg\phi})) \rightarrow \mathrm{K}(\overline{\neg\phi})$	by (I)
8.	$\vdash \mathrm{K}(\overline{\neg\phi})$	from $5, 6, 7$
9.	$\vdash \phi$	from 8 and 1, contra 4

Remarks. The paradox has often been discussed together with the question whether certain explications of 'propositions' are adequate, see e.g. Asher and Kamp [18] and Thomason [406]. Several authors investigated the relation of the Knower Paradox to the principle of epistemic closure, e.g. Maitzen [241], Cross [87, 88], Uzquiano [437], Égré [114]. Restricting or avoiding epistemic closure is, however, only one of several approaches to the Knower Paradox. For example, Lee [231] assumes that our intuitive notion of knowledge is circular. Recently, Égré [114] thoroughly examined two readings of formulas in the paradox (metalinguistic and self-referential reading) and offered a syntactical-hierarchical solution using provability interpretation of modal logic. Dean and Kurokawa [93] examined proof-theoretic evidence for statements containing iterated K and found the culprit of the contradiction there.

Anderson [8] solved the paradox using a hierarchy of K-predicates, i.e. a certain variant of Tarski's hierarchy of T-predicates or Russell's RTT. Although he did not consider the result entirely elucidating, he viewed the idea of enriching EL using a hierarchy of knowledge operators as promising:

> I don't suppose that I have proved that knowledge has levels, or even explained exactly what this means. Much more technical and philosophical work needs to be done. But the intuitive ideas have considerable appeal as a solution to the Knower Paradox and an independent motivation suggested by Gödel's Theorems. Epistemic logic may have an interesting future after all. Up the hierarchy!
>
> Anderson [8], p. 354

On the other hand, e.g. Grim (e.g. [160, 159]) and Tucker [430] neglected the hierarchical approach to the Knower Paradox, since it thwarts an attempt to build a 'global' notion of truth or knowledge (see 7.3 above). Yet we have seen that no improvement concerning e.g. the T-predicate can be expected given Tarski's undefinability theorem. With regards to K, if we individuated it as applicable to 'propositions' that are explicated as certain constructions of TTT (for reasons see 1, 4, 6), no 'global' operator K applicable to all constructions is possible.

Remark. The authors do not address hyperintentensional conceptions by e.g. Tichý [422] or Moschovakis [265]. Moreover, it seems that Tucker [430] identifies the hyperintensional conception of meaning with sententialism (which he rightly criticises).

To solve the paradox, the very predicate "K" applicable to sentences (formulas) could be refuted.[16] Strictly speaking, any sentence is a mere sequence of letters; attitudes explicated as relations to such sequences of letters are obviously inadequate as a model of belief: they are criticisable similarly as the original kind of sententialism (1.2.1). One would thus reject the axioms (1) – (3) by considering them (and thus the whole paradoxical inference) irrelevant to knowledge.

But the paradox can be reformulated, using e.g. the construction $[\mathbf{K}^k\, x\, {}^0C^k]_{tw}$ instead of the formula $\mathrm{K}(\overline{\varphi})$. Again, there arises a question where the root of contradiction is. Though I doubt the naive law of factivity of K, i.e. (T), I consider rather the law (I) to be the culprit of the paradox. I thus join the club of the critics of the principle of epistemic closure.

The unacceptability of (I) can be shown if using two v-congruent constructions such as $[[1+2] =^{\rho} 3]$ and *FLT*. Because of their v-congruence, *FLT* is

[16] This is not to say that no alethic modality can be treated as applicable to expressions. For example, $\Box$ can be well treated as a predicate applicable to sentences (see e.g. Halbach, Leitgeb and Welch [163]), although I would add that such syntactical treatment of modality should be relativised to a language, similarly as the T-predicate applicable to sentences.

derivable from $[[1+2] =^{\rho} 3]$ but it is unjustifiable to assume that if an agent knows $[[1+2] =^{\rho} 3]$ then she also knows *FLT*. The agent's inference is bound by limits described in the previous chapter as derivation w.r.t. derivation systems (6.2.1).

Now consider that one of the constructions manipulated by the law (I) is $\boxed{K^1}$. This valid construction can be inferred from another valid construction, but it is unjustifiable to assume that an agent knows $\boxed{K^1}$: $\boxed{K^1}$ cannot be known^1 (see 7.5). If we place $\boxed{K^1}$ instead of $\neg\phi$ in the paradoxical inference,[17] the inference runs correctly until the application of (I), which yields the invalid construction $[\mathbf{K}^1\, x^0 \boxed{K^1}]_{tw}$.

7.7 Church-Fitch's knowability paradox

The famous *Church-Fitch's knowability paradox* (*FP*) seems to prove that the epistemic optimism known as *verificationism*, according to which every truth is knowable, is wrong.

The FP was discovered by Church in 1945 when reviewing Fitch's paper (this anonymous review was recently identified and published as [74]), while it was published by Fitch [125]. For a long time the paradox remained nearly unnoticed, but it has attracted a considerable amount of attention recently.[18]

Some approaches attempt to solve the FP by restricting a. the formulation of verificationism, while some other approaches b. revise its underlying logic. The *typing knowledge* (*TK*) combines a. and b., for the stratification of the K-operator into an infinite hierarchy of the operators K^1, K^2, ..., K^n (while each has a restricted range of applicability) revises both rules and 'propositions' used in the FP.

The typing approach was implicitly considered by Church [74]:

> Of course the foregoing refutation of Fitch's definition of value is strongly suggestive of the paradox of the liar and other epistemological paradoxes. It may be therefore that Fitch can meet this particular objection by incorporating into the system of his paper one of the standard devices for avoiding the epistemological paradoxes.
>
> Church [74], p. 17

but Williamson [449] showed its first explicit application. He in fact used the *Tarskian TK* (as I will call it), since he evoked the stratification of T-predicate derivable from Tarski's seminal paper [398], while he did not mention Russell

17 $\boxed{K^1}$ cannot be placed in lieu of ϕ because the consequent of $\phi \to \neg\phi$ cannot be invalid.

18 See the entry by Brogaard and Salerno [49] for an overview. Most of my results concerning the FP were published in Raclavský [325, 331, 336].

or any Russellian topic. Such TK has been adopted by many writers: Paseau [290], Halbach [162], Paseau [291],[19] Florio and Murzi [131], Jago [186], Carrara and Fassio [59].

On the other hand, Linsky [239] applied the *Russellian TK* to the FP; see also Giaretta [155] who focused on the relationship of such a solution to the Russellian logical framework. They both used (a simplification of) Church's [66] r-types (see 2.2). Justification of RTT is rather rudimentary in Linsky's paper.

Most of the just mentioned writers, notably Carrara and Fassio [59], consequently objected that the typing approach is *ad hoc*, for it has no paradox-independent motivation. However, if one clearly distinguishes the Tarskian and the Russellian TK, and embraces rather the latter, such an objection becomes unfounded, because the Russellian TK relies on non-circular forming of 'propositions', which is governed by VCP, a principle independent on paradoxes.

On the other hand, the Tarskian TK seems to be motivated merely by prevention of paradoxes, and so it seems to have no independent motivation. I am nevertheless convinced that such an objection disregards an important argument in favour of the Tarskian approach: the primary purpose of typing was to help delivering a formally correct explication of the intuitive notions, e.g. truth; since the approach is successful in this respect, philosophical 'worries' should be sidelined.[20]

In the next few sections, I show that (Russellian) TK solves the FP (7.7.1) and discuss three possible objections to it. I also solve its revenge form (7.7.3) that was proposed by Williamson [449]. Finally (7.7.4), I demonstrate that the typing approach cannot avoid the FP (or its revenge form), if one admits the GPR (2.4.4).

7.7.1 The typing solution to the FP

The FP is generated by a combination of apparently acceptable principles and some unproblematic rules of first-order logic, EL and modal logic. The FP begins with the evident principle that we are not omniscient:

(NonOmn) $\quad \exists\varphi(\varphi \wedge \neg K\varphi) \quad$ *the Principle of Non-omniscience*

to which the *Principle of Verificationism* (or *Knowability*) is added:

[19] Halbach [162] suggested a paradox similar to the FP, yet Paseau [291] noticed its closer similarity to the Knower paradox (see 7.6); see also Rosenblatt [351].

[20] Various further philosophical objections targeting material adequacy of TK are discussed by Paseau [290].

(Ver) $\forall\varphi(\varphi \to \Diamond K\varphi)$ *the Principle of Verificationism*

But FP's inference paradoxically achieves

(Omn) $\forall\varphi(\varphi \to K\varphi)$ *the Principle of Omniscience*

The FP deploys the unrestricted operator K and the unrestricted (meta-) variable φ. Note that both ideas contradict VCP.

Example 78 (The untyped FP)

1′.	$\exists\varphi(\varphi \wedge \neg K\varphi)$	(NonOmn)
2′.	$\varphi \wedge \neg K\varphi$	from 1′ by ($\exists$-E)
3′.	$\forall\varphi(\varphi \to \Diamond K\varphi)$	(Ver)
4′.	$(\varphi \wedge \neg K\varphi) \to \Diamond K(\varphi \wedge \neg K\varphi)$	from 3′ by (UI) (subst. of 2′ for φ)
5′.	$\Diamond K(\varphi \wedge \neg K\varphi)$	from 4′ and 2′ by (MP)
6′.	$K(\varphi \wedge \neg K\varphi)$	assumption *per absurdum*
7′.	$K\varphi \wedge K\neg K\varphi$	from 6′ by (Dist)
8′.	$K\varphi \wedge \neg K\varphi$	from 7′ by (Fact)
9′.	$\neg K(\varphi \wedge \neg K\varphi)$	*reductio* (for 8′ is a contradiction)
10′.	$\Box\neg K(\varphi \wedge \neg K\varphi)$	from 9′ by (Nec)
11′.	$\neg\Diamond K(\varphi \wedge \neg K\varphi)$	from 10′ by (ER)

Here 11′ contradicts 5′, which means that the addition of (Ver) to (NonOmn) leads to (Omn).

TK properly 'types' the rules and 'propositions' of the FP with accordance with VCP. The *reductio* part is then prevented. To make it more comprehensible, I employ the following (schematic) notational conventions, while I always intend constructions/rules that repeatedly occur in this book, esp. chapter 6.[21]

[21] In the abbreviated names of the rules (see the next convention), "k" does not always match the order of the constructions involved in them.

Notational convention 38 (Simplified records of constructions for the FP)

Abbreviation	*for ...*
φ^k, ψ^k, etc.	(a) ${}^2c^k, {}^2d^k$, etc., if it does not occur in the immediate scope of $\mathbf{K}^k$ or $\mathbf{Tr}^k$, but e.g. near to ${}^{(0)}\wedge$ (in this case, I assume that partiality is amended) (b) c^k, d^k, etc., if it occurs in the immediate scope of $\mathbf{K}^k$ or $\mathbf{Tr}^k$
f_x	$[f^{(\pi\tau)}\, x^\tau]_{tw}$
$\mathbf{K}^k\varphi^k, \mathbf{K}^k_x\varphi^k$, etc.	$[\mathbf{K}^k\, x\, \varphi^k]_{tw}$, etc.
$\Box\varphi^k$, $\Diamond\varphi^k$	$[\underset{\pi}{\Box}[\lambda t\lambda w.\varphi^k]]$, $[\underset{\pi}{\Diamond}[\lambda t\lambda w.\varphi^k]]$ (to avoid ascent of orders when using $\Box^k$ and $\Diamond^k$)
$\forall x(...)$, $\exists x(...)$	$[\forall^\tau[\lambda x^\tau[...]]]$, $[\exists^\tau[\lambda x^\tau[...]]]$ ("τ" is just the numeral "k", if τ is $*_k$)
$C_{3(C_1/C_2)}$	${}^2[\mathbf{Sub}^k\, {}^0C_1\, {}^0C_2\, {}^0C_3]$ ("(C_1/C_2)" has the smallest conceivable scope)

Notational convention 39 (Typed rules of the FP)

(NonOmn^k)	$\exists\varphi^k(\varphi^k \wedge \neg\mathbf{K}^k\varphi^k)$	*the Principle of Non-omniscience*
(Ver^k)	$\forall\varphi^k(\varphi^k \rightarrow \Diamond\mathbf{K}^k\varphi^k)$	*the Principle of Verificationism*
(MP^k)	$\varphi^k\,;\ \varphi^k \rightarrow \psi^k \vdash \psi^k$	*Modus Ponens* (a variant)
(UI^k)	$\forall x\varphi^k \vdash \varphi^k_{(t/x)}$	*Universal Instantiation*
($\exists$-E^k)	$\exists x\varphi^k \vdash \varphi^k_{(t/x)}$	*$\exists$-Elimination* (condition: t is 'fresh')
(Fact^k)	$\mathbf{K}^k\varphi^k \vdash \varphi^k$	*the Factivity Rule*
(Dist^k)	$\mathbf{K}^k(\varphi^{k-1} \wedge \psi^k) \vdash \mathbf{K}^k\varphi^{k-1} \wedge \mathbf{K}^k\psi^k$	*the Rule of Distributivity of Knowledge* (a specific variant)
(Nec^k)	if $\vdash \varphi^k$, then $\vdash \Box\varphi^k$	*the Rule of Necessitation*
(ER^k)	$\neg\Diamond\varphi^k \dashv\vdash \Box\neg\varphi^k$	*the Exchange Rule for Mod. Operators*

Example 79 (The typed (part of) the FP)

1.	$\exists\varphi^1(\varphi^1 \wedge \neg\mathbf{K}^1\varphi^1)$	(NonOmn^1)
2.	$\varphi^1 \wedge \neg\mathbf{K}^1\varphi^1$	from 1 by ($\exists$-E^2)
3.	$\forall\varphi^2(\varphi^2 \rightarrow \Diamond\mathbf{K}^2\varphi^2)$	(Ver^2)
4.	$(\varphi^1\wedge\neg\mathbf{K}^1\varphi^1) \rightarrow \Diamond\mathbf{K}^2(\varphi^1\wedge\neg\mathbf{K}^1\varphi^1)$	from 3 by (UI^2) (subst. of 2 for φ^2)
5.	$\Diamond\mathbf{K}^2(\varphi^1 \wedge \neg\mathbf{K}^1\varphi^1)$	from 4 and 2 by (MP^2)
6.	$\mathbf{K}^2(\varphi^1 \wedge \neg\mathbf{K}^1\varphi^1)$	assumption *per absurdum*
7.	$\mathbf{K}^2\varphi^1 \wedge \mathbf{K}^2\neg\mathbf{K}^1\varphi^1$	from 6 by (Dist^2)
8.	$\mathbf{K}^2\varphi^1 \wedge \neg\mathbf{K}^1\varphi^1$	from 7 by (Fact^2)

Unlike in its untyped version, the *reductio* is prevented, since step 8 is apparently not a contradiction (but see the discussion in point a. in the next (sub)section).

7.7.2 Three inconclusive objections to the typing solution

In this section, I am going to reject three further objections (a.–c.) to the typing solution to the FP.

a. The typing solution to the FP requires that the rule

$$\mathbf{K}^2\varphi^1 \vdash \mathbf{K}^1\varphi^1$$

is considered invalid, otherwise the proposition 8 would be contradictory (Williamson [449]). Nevertheless, mere principles of type-theoretic correctness provide no reason for its adoption or rejection.

The authors writing about the FP attempted to explain the invalidity of the rule e.g. by appealling to 'epistemic access' to φ^1, see e.g. Williamson [449], Linsky [239], Paseau [290, 291]. Linsky also cautiously wrote that 'propositions' are typed due to their content, while logical relations between them reflect procedures for determination of epistemological states.

I suggest to generalising Linsky's idea, evoking somewhat Plato's classical 'JTB theory' of knowledge: for knowing^m φ^k, one needs justification^m of φ^k by a certain reason^m (an mth-order o-construction), for $k \leqslant m$ (where $\mathbf{J}^m$ (to be justified) $/(\pi *_k)$; $\mathbf{R}^m$ (to be a reason for) $/(\pi *_m *_k)$).[22]

$$\vDash \mathbf{J}^m\varphi^k \Leftrightarrow_o \exists\psi^m\mathbf{R}^m(\psi^m, \varphi^k) \quad \text{for } k \leqslant m$$

The justification surely constitutes the attitude towards φ^k being knowledge, rather than mere belief or contemplation.

Remark. The reasons φ^ms serving for justification of φ^k can be, say, inevitable steps in a derivation of φ^k. Some ψ^ms can be epistemic 'propositions' or 'propositions' about an epistemic route to φ^k – these can be e.g. mth-order 'propositions' certifying that φ^ks were acquired from a reliable source. But one can leave the exact nature of ψ^ms undecided and so not rely exclusively on the 'epistemic explanation' of the invalidity of the rule in question. (Cf. also the discussion in 7.7.4.)

The considerations entail that for knowing^2 the 'proposition' φ^1 one needs a certain ψ^2 that helps to justify^2 φ^1. The respective reason^2 thus makes $\mathbf{K}^2\varphi^1$ irreducible to $\mathbf{K}^1\varphi^1$. The rule $\mathbf{K}^2\varphi^1 \vdash \mathbf{K}^1\varphi^1$ is therefore invalid.

---

[22] Recall that "justification" is a buzzword in recent EL, see e.g. Artemov and Kuznets [14] or Baltag, Renne and Smets [22].

b. Such a conclusion even leads to the rejection of the argument raised by Jago [186], but his argument can be refuted in a more straightforward manner. Jago assumed that the proponent of TK accepts the principle (notation adjusted)

$$(\mathrm{KT}^2) \quad K^1\varphi^1 \leftrightarrow K^2T^1\varphi^1$$

where T^1 operates on typed 'propositions'. Here are the four crucial steps in Jago's modification of the FP:

$1^j.\ K^2(T^1\varphi^1 \wedge \neg K^1\varphi^1)$	assumption
$2^j.\ K^2T^1\varphi^1 \wedge K^2\neg K^1\varphi^1$	from 1^j. by (Dist^2)
$3^j.\ K^1\varphi^1 \wedge K^2\neg K^1\varphi^1$	from 2^j. by (KT^2)
$4^j.\ K^1\varphi^1 \wedge \neg K^1\varphi^1$	from 3^j. by (Fact^2)

Nevertheless, I reveal that $T^1\varphi^1$ is clearly υ-congruent with φ^1 by the generally accepted T-axiom $T^1\varphi^1 \leftrightarrow \varphi^1$. (KT$^{2(2)}$) is thus υ-congruent with

$$K^1\varphi^1 \leftrightarrow K^2\varphi^1$$

which can hardly be maintained by any proponent of TK (see e.g. the considerations in a.), and so she easily contests Jago's version of the FP.

c. Carrara and Fassio [59] raised an argument attempting to show that the typing approach to the FP is internally incoherent. They argued that, instead of 7, the typing approach must admit (notation adjusted)

$$K^1\varphi^1 \wedge K^2\neg K^1\varphi^1,$$

for $K^2\varphi^1$ is (allegedly) not type-theoretically possible. Using (Fact^2), one then derives the contradictory formula $K^1\varphi^1 \wedge \neg K^1\varphi^1$, and so the *reductio* is not blocked.

However, the typing approach framed within standard RTTs (TTT included) is not affected by the argument because it implements the PCC (2.4.4). So the approach embraces 'propositions' such as $\mathbf{K}^2\varphi^1$, in consequence of which there is no obligation to admit $\mathbf{K}^1\varphi^1 \wedge \mathbf{K}^2\neg\mathbf{K}^1p^1$ instead of 7. In other words, the argument is only succesful against TK framed within non-cumulative RTTs.[23]

7.7.3 The typing solution to the Revenge Form of the FP

As mentioned in the introductory section, Williamson [449] first suggested the *Revenge Form of FP for the Typing Approach* (*RFP*):

[23] Consequences of acceptance of cumulativity for the solution of paradoxes were studied by Peressini [292]; the present case is not discussed there.

> We seem able to grasp the idea that φ is *totally unknown*, in a sense which entails that φ is unknown$_i$ for each level i, but which does not entail that φ is untrue. If so, we can simply adapt Fitch's argument by considering the proposition that φ is totally unknown truth, since that proposition cannot be known$_i$ for any level i. Naturally, such quantification over levels must be handled with great care
>
> Williamson [449], p. 281

The inference considered by Williamson, and mentioned by Carrara and Fassio [59], was further elaborated by Hart [167]. Hart explicitly stated its conclusion that every truth is necessarily known at some *type-level t* (this notion has never been exactly defined).

Here I formally reconstruct Hart's intended inference (the key formula "$\forall t \neg \mathrm{K}^t p$" was proposed by Carrara and Fassio [59]). It does not exactly parallel the inference of the FP, but it is fairly similar.

Example 80 (The Revenge Form of the FP for Typing Approach (RFP))

1^r.	$\varphi \wedge \forall t \neg K^t \varphi$	assumption (of order $t+1$) *per absurdum*
2^r.	$\Diamond K^{t+1}(\varphi \wedge \forall t \neg K^t \varphi)$	from 1^r. by (Ver)
3^r.	$\Diamond (K^{t+1}\varphi \wedge K^{t+1}\forall t \neg K^t \varphi)$	from 2^r. by (Dist)
4^r.	$\Diamond (K^{t+1}\varphi \wedge \forall t \neg K^t \varphi)$	from 3^r. by (Fact)
5^r.	$\Diamond (K^{t+1}\varphi \wedge \neg K^{t+1} \varphi)$	from 4^r. by (UI)
6^r.	$\neg(\varphi \wedge \forall t \neg K^t \varphi)$	*reductio* (since 5^r. is a contradiction)
7^r.	$\Box\neg(\varphi \wedge \forall t \neg K^t \varphi)$	from 6^r. by (Nec)
8^r.	$\Box(\varphi \rightarrow \neg\forall t \neg K^t \varphi)$	from 7^r. by classical logic
9^r.	$\Box(\varphi \rightarrow \exists t K^t \varphi)$	from 8^r. by De Morgan Law
10^r.	$\Box\forall\varphi(\varphi \rightarrow \exists t K^t \varphi)$	from 9^r. by the Universal Generalisation

But 10^r contradicts our assumption that there are unknown truths.

However, the inference is suspicious from the very beginning. Consider the formula "$\forall t \neg K^t \varphi$" first: if the use of $\forall$ is not hollow, $\forall$ binds some variable within "$\neg K^t \varphi$". Evidently, that variable is t. Yet holding such an assumption, one departs from the official notation of RTT in which "K^k" (or "$\mathbf{K}^k$") is a primitive symbol, so the numeral "k" is its irreplaceable part, and so it is not a variable.

To challenge the (Russellian) TK one must therefore accept that, since formulas standing for 'propositions' have to be typed,

> "$K^t\varphi$" is in fact short for "$\lambda\varphi^k t.\mathbf{K}'^m(\varphi^k, t)$" (for $1 \leq k \leqslant m$), where $\mathbf{K}'^m$ is a novel binary operator of type $(\pi\iota *_k \rho)$.

Of course, there are several possibilities of how to interpret "t" and, consequently, several possibilities as to how to understand "$\mathbf{K}'^{m}$". On the most probable reading, however, t ranges over natural numbers $\mathbb{N}^{+}$ that represent orders, as suggested. For simplification of considerations assume $m = k$. (It can be easily checked that my findings will not hinge on the two particular assumptions.)

On the preferred reading, RFP's second 'proposition' is

$$2^{r'}.\ \Diamond\mathbf{K}'^{k+1}(\varphi^{k} \wedge \forall\lambda t\neg\mathbf{K}'^{k}(\varphi^{k}, t), t+1)$$

Applying appropriate rules of distributivity and factivity that govern these two novel operators $\mathbf{K}'^{k}$ and $\mathbf{K}'^{k+1}$ one infers

$$5^{r'}.\ \Diamond(\mathbf{K}'^{k+1}(\varphi^{k}, t+1) \wedge (\forall t\neg\mathbf{K}'^{k}(\varphi^{k}, t), t+1))$$

Applying (UI) to the right conjunct of $5^{r'}$. we get

$$6^{r'}.\ \Diamond(\mathbf{K}'^{k+1}(\varphi^{k}, t+1) \wedge \neg\mathbf{K}'^{k}(\varphi^{k}, t+1)),$$

which is not a contradiction. Consequently, the *reductio* part of the RFP is prevented, similarly as in the typed FP.

So the RFP is based on a hidden equivocation, too. If the (seemingly correctly typed) formulas of the inference are disambiguated and properly typed, and thus matched with some existing 'propositions', the inference does not go through. (However, thanks to the GPR, it can be adjusted to an unsolvable paradoxical inference, as the FP in the next (sub)section.)

7.7.4 Reducibility and the typing solution to the FP

In this section, I show that TTT does not avoid the FP (or the RFP), if the GPR (2.4.4) is accepted. For my investigation, a somewhat narrower principle than the GPR is sufficient: for every $(m+1)$st-order knowledge operator, e.g. $\mathbf{K}^{m}$, operating on kth-order entities, for $1 \leqslant k \leqslant m$, there exists a $(k+1)$st-order v-congruent one, e.g. $\mathbf{K}^{k}$.

Note that the GPR does not say that for every $(m+1)$st-order operator there exists a $(k+1)$st-order congruent one, for any $1 < k$. Because if such a principle were valid, the whole hierarchy of operators would collapse to its bottom order 1, becoming thus pointless. Note also that, given the PCC, the GPR sounds particularly trivial, which I am going to illustrate in my first example (i) below. Nevertheless, the GPR says something distinct from the PCC, and for that my second example (ii) is convenient.

(i) To speak about $(m+1)$st-order operator*s* of knowledge is correct. By the PCC, the 2nd-order operator $\mathbf{K}^{1}$ is also of order 3, 4, etc., while there is also a novel, 3rd-order operator $\mathbf{K}^{2}$ that has distinct (richer) range of extensions, and is also of order 4, 5, etc. (Etc.) Thus, for example, the 4th-order operators $\mathbf{K}^1$ and $\mathbf{K}^2$ are each v-congruent with a certain 3rd-operator, namely $\mathbf{K}^1$ and $\mathbf{K}^2$, respectively, and so the GPR is trivially satisfied.

(ii) Now consider a 1st-order 'proposition' φ^1 and the $(k+1)$st-order operator $\mathbf{K}^k$ (for $1 \leqslant k$) that operates on φ^1s. There exists a 2nd-order v-congruent operator I will denote "$\mathbf{K}^{[k]1}$" to distinguish it from $\mathbf{K}^1$ (the two operators are not identical):

Convention 40 (Lower-order K-operator corresponding to a higher-order one, $\mathbf{K}^{[k]1}$)

"$\mathbf{K}^{[k]1}$", for $1 < k$, stands for a 2nd-order operator that is v-congruent with $\mathbf{K}^k$.

Below, I will use two rules governing pairs of such v-congruent knowledge operators. The first rule states v-congruence of $\mathbf{K}^{[2]1}\varphi^1$ with $\mathbf{K}^2\varphi^1$, and so it expresses the GPR for this particular case.

Definition 105 (The Rule ($\mathbf{K}^2 \Leftrightarrow \mathbf{K}^{[2]1}$))

$\mathbf{K}^2\varphi^1 \dashv\vdash \mathbf{K}^{[2]1}\varphi^1$

Note that the rule does not license us to claim v-congruence of $\mathbf{K}^{[2]1}$ with $\mathbf{K}^1$. On the other hand, the following rule and its inverse form seem valid: if one knows^1 φ^1, φ^1 is also known$^{[2]1}$, since $\mathbf{K}^1$ serves as a 2nd-order operator to which there corresponds a certain $\mathbf{K}^{[2]1}$.

Definition 106 (The Rules ($\mathbf{K}^1 \Rightarrow \mathbf{K}^{[2]1}$) and ($\neg\mathbf{K}^{[2]1} \Rightarrow \neg\mathbf{K}^1$))

$\mathbf{K}^1\varphi^1 \vdash \mathbf{K}^{[2]1}\varphi^1$

$\neg\mathbf{K}^{[2]1}\varphi^1 \vdash \neg\mathbf{K}^1\varphi^1$

The next inference starts with a bit more general version of (NonOmn^1), from which (NonOmn^1) is nevertheless derivable using ($\neg\mathbf{K}^{[2]1} \Rightarrow \neg\mathbf{K}^1$).

Example 81 (Typed Form of the FP that assumes the RP)

1^r.	$\exists\varphi^1(\varphi^1 \wedge \neg\mathbf{K}^{[2]1}\varphi^1)$	(NonOmn$^{[2]1}$)
2^r.	$\varphi^1 \wedge \neg\mathbf{K}^{[2]1}\varphi^1$	from 1^r by ($\exists$-E^1)
3.	$\forall\varphi^2(\varphi^2 \rightarrow \Diamond\mathbf{K}^2\varphi^2)$	(Ver^2)
4^r.	$(\varphi^1 \wedge \neg\mathbf{K}^{[2]1}\varphi^1) \rightarrow \Diamond\mathbf{K}^2(\varphi^1 \wedge \neg\mathbf{K}^{[2]1}\varphi^1)$	from 3 by (UI^2) (subst. of 2^r for φ^2)
5^r.	$\Diamond\mathbf{K}^2(\varphi^1 \wedge \neg\mathbf{K}^{[2]1}\varphi^1)$	from 4^r and 2^r by (MP^2)
6^r.	$\mathbf{K}^2(\varphi^1 \wedge \neg\mathbf{K}^{[2]1}\varphi^1)$	assumption *per absurdum*
7^r.	$\mathbf{K}^2\varphi^1 \wedge \mathbf{K}^2\neg\mathbf{K}^{[2]1}\varphi^1$	from 6^r by (Dist^2)
8^r.	$\mathbf{K}^2\varphi^1 \wedge \neg\mathbf{K}^{[2]1}\varphi^1$	from 7^r by (Fact^2)
8^{r2}.	$\mathbf{K}^{[2]1}\varphi^1 \wedge \neg\mathbf{K}^{[2]1}\varphi^1$	from 8^r. by ($\mathbf{K}^2 \Leftrightarrow \mathbf{K}^{[2]1}$)

9^r.	$\neg\mathbf{K}^2(\varphi^1 \wedge \neg\mathbf{K}^{[2]1}\varphi^1)$	*reductio*
10^r.	$\Box\neg\mathbf{K}^2(\varphi^1 \wedge \neg\mathbf{K}^{[2]1}\varphi^1)$	from 9^r by (Nec$^\pi$)
11^r.	$\neg\Diamond\mathbf{K}^2(\varphi^1 \wedge \neg\mathbf{K}^{[2]1}\varphi^1)$	from 10^r by (ER$^\pi$)

Hence, 11^r is contradictory to 5^r, which means that adding (Ver^2) to (NonOmn$^{[2]1}$) leads to (Omn$^{[2]1}$), i.e. $\forall\varphi^1(\varphi^1 \rightarrow \mathbf{K}^{[2]1}\varphi^1)$.

TK that adopts the GPR thus cannot avoid such a paradox. Similarly for the GPR-version of the RFP.

Comment on verificationism.

That TK does not block the above paradox should lead us to rethink the role of TK as a guardian of verificationism, as it was assumed to be by critics of the typing solution to the FP. Strictly speaking, TK is independent of the goals of verificationism: one can adhere to the former without accepting the latter (and *vice versa*).

But I even believe that adoption of TK for protection of verificationism is undermined from the start. Recall that, if properly typed, (Ver^k) only says that kth-order truths are knowable^k. This particular variant of verificationism thus omits all truths that cannot be in the scope of $\mathbf{K}^k$.

True, thanks to the GPR, the expressive strength of the thesis can surely be increased as follows:

$$\exists\underline{\mathbf{K}^{k+1}}\forall\lambda\varphi^k(\varphi^k \rightarrow \Diamond\underline{\mathbf{K}^{k+1}}\varphi^k)$$

where $\underline{\mathbf{K}^{k+1}}$ is a variable for $(k+1)$st-order operators of knowledge, i.e. $\mathbf{K}^k$, $\mathbf{K}^{[k+1]k}$, $\mathbf{K}^{[k+2]k}$, etc. Nevertheless, they will always remain truths that are outside of the range of the thesis. As VCP tells us, there are always some truths beyond the reach of any general 'proposition' one will never be capable of knowing. It thus seems that to give up verificationism is the best solution to the FP.

Appendix A

The Proof of the Compensation Principle

As noted in Raclavský [328] (see also 2), Tichý [422] apparently overlooked that executions ${}^1C$ and ${}^2C$ may contain free variables even if C does not contain a free variable. The omission is visible in his definition of the notions of free variable and construction's rank. Consequently, his Compensation Principle (which uses the 'metalinguistic' substitution function) and its proof has no straightforward generalisation for constructions of order $1 < n$, since the aforementioned executions are higher-order ones.

Below, I offer a novel definition of the construction's rank (adopted from Raclavský, Kuchyňka and Pezlar [339]) and adequate proof of the generalised version of the Compensation Principle. The proof of the theorem is more than two times longer than Tichý's proof (which only consists of sentences which I denote (1), (2), (r_1), (r_{+1}.i), (r_{+1}.ii)).

Definition 107 (Rank, r)

Let C be a construction of order n.

1. The *rank* of C is 1, if C is of the form

   x, ${}^0X$, ${}^1X$, ${}^2X$

   where X is not a construction.

2. The *rank* of C is $r + 1$, if C is of the form

   (a) ${}^0C_0$, ${}^1C_0$, ${}^2C_0$, $[\lambda x_1...x_m.C_0]$ and r is the rank of C_0

   (b) $[C_0\, C_1...C_m]$ and r is the greatest among the ranks of $C_0, C_1, ..., C_m$

   where $C_0, C_1, ..., C_m$ are constructions of order n in the case of $[\lambda x_1...x_m.C_0]$ and $[C_0\, C_1...C_m]$, or of order $n-1$, for $1 \leqslant (n-1)$, in the case of ${}^0C_0$, ${}^1C_0$, ${}^2C_0$.

The proof of the Compensation Principle requires three auxiliary theorems. The first one, the 'Replacement of Free Variables' theorem, was proved by Tichý [422] for constructions of order 1 (not n), and is rather similar to his proof of the Compensation Principle. Let C be what is v-constructed by C (similarly for D, x etc.).

Theorem 10 (Replacement of Free Variables)

Let C be a construction of order n, $x, ..., x_m$ distinct variables of the same respective types and such that for $1 \leqslant i \leqslant m$, either y_i is the same as x_i or it does not occur in C. Then C $v(\mathrm{D}_1/x_1, ..., \mathrm{D}_m/x_m)$-constructs C iff $C_{(y_1/x_1,...,y_m/x_m)}$ $v(\mathrm{D}_1/y_1, ..., \mathrm{D}_m/y_m)$-constructs C.

Remark. The theorem utilises the fact that when proving that, for any v, two constructions C_1 and C_2 are v-congruent, it does not matter if we consider an arbitrary v evaluating C_i's free variable x or its appropriate replacement y. Tichý's [422] original proof requires analogous amendment as is suggested in the proof of the Compensation Principle below.

The other two auxiliary theorems are rather obvious.[1]

Theorem 11 **('$^{20}C \cong C$')**

For any v and construction C, $^{20}C \cong C$.

Proof. By the specification of constructions of the form ^{0}C and ^{2}C if C v-constructs C, then ^{20}C v-constructs as follows: (i) it makes ^{0}C to v-construct C as such and then, (ii) it makes C to v-construct C. Similarly for the case when C v-constructs _, i.e. when C and thus also ^{20}C are v-improper.

Theorem 12 **('$^{1}C \cong C$')**

For any v and construction C, $^{1}C \cong C$.

Proof. Trivial, given the specification of constructions of the form ^{1}C.

Theorem 1 (The Compensation Principle)

Let C be any construction of order n. For any valuation v and any construction D of order n, if D v-constructs D, then $C_{(D/x)}$ v-constructs C iff C $v(\mathrm{D}/x)$-constructs C.

---

[1] See Raclavský [328]; cf. also Duží, Jespersen and Materna [110]. In Raclavský, Kuchyňka and Pezlar [339], Kuchyňka proved interderivability of sequents whose succedents are ^{20}C and C.

Proof. (1) If x is not free in C, then $C_{(D/x)}$ is C and the condition of the Theorem is obviously satisfied. (2) Assume, therefore, that x is free in C.

(r_1) If C is of rank 1, then C is x, which $v(\mathrm{D}/x)$-constructs D. Since $C_{(D/x)}$, which is $x_{(D/x)}$, is D, it v-constructs D (by hypothesis), and so the Theorem is satisfied.

(r_2) If C is of rank 2, then C is of the form (i) $[\lambda x_1...x_m.C_0]$, (ii) ${}^1C_0$, (iii) ${}^2C_0$, or (iv) $[C_0\, C_1...C_m]$, where for $0 \leqslant i \leqslant m$, some C_i is x.

In case (i), C is $[\lambda x_1...x_m.x]$. Let C $v(\mathrm{D}/x)$-construct C and $C_{(D/x)}$, which is $[\lambda x_1...x_m.x_{(D/x)}]$, v-constructs C′. It is easy to show that C′ is C and so the Theorem is satisfied. Since x $v(\mathrm{D}/x)$-constructs D iff $x_{(D/x)}$ v-constructs D (cf. r_1), both C′ and C are one and the same function that maps $\langle \mathrm{X}_1, ..., \mathrm{X}_m \rangle$ to D.

In case (ii), C is ${}^1x$. By the theorem '${}^1C \cong C$', ${}^1x$ is v-congruent with x and, given (r_1), it is easy to see that the Theorem is satisfied.

In case (iii), C is ${}^2x$. If D v-constructs D that is not a v-proper construction, then ${}^2x$ is $v(\mathrm{D}/x)$-improper and ${}^2C_{0(D/x)}$, which is ${}^2x_{(D/x)}$, i.e. ${}^2D$, is also v-improper; so the Theorem is satisfied. If, on the other hand, D does v-construct a v-proper construction D, then ${}^2x$ $v(\mathrm{D}/x)$-constructs what is v-constructed by D, call that object X. Since ${}^2C_{0(D/x)}$, which is ${}^2x_{(D/x)}$, is ${}^2D$, it v-constructs what is v-constructed by D, namely X, so the Theorem is satisfied, too.

In case (iv), C is $[C_0\, C_1...C_m]$, where some C_i is x. The other subconstructions of C comprise (a) variable(s) distinct from x or construction(s) of the form (b) ${}^0X$ (with no free x) or of the form (c) ${}^1X$ or ${}^2X$, where X is a variable or X is not a construction. Given (r_1), it is easy to see that the Theorem is satisfied in cases when C contains subconstructions of type (a), (b), but even (c) (cf. case (ii) above), or their combinations. (Some of Cs and $C_{(D/x)}$s may not be v-proper – e.g. for the reason that the function C_0 v-constructed by C_0 is applied to an argument $\langle \mathrm{C}_1, ..., \mathrm{C}_m \rangle$ for which it is undefined, or because the type of C_0 does not rightly match with the types of $\mathrm{C}_1, ..., \mathrm{C}_m$, or because some C_i is v-improper and the function C_0 (if any) does not receive a suitable argument. Details are left to the reader.)

(r_{+1}) Now assume, as an induction hypothesis, that the Theorem is satisfied by any construction C of rank $2 \leqslant r$, and consider a C of rank $r+1$ (i.e. of rank greater than, or equal to, 3). C is of the form (i) $[C_0\, C_1...C_m]$, (ii) $[\lambda x_1...x_m.C_0]$, (iii) ${}^1C_0$, or (iv) ${}^2C_0$.

In case (i), C is of the form $[C_0\, C_1...C_m]$ and it is easy to see that the Theorem is satisfied.

In case (ii), C is of the form $[\lambda x_1...x_m.C_0]$, and, by the definition of SUB, $C_{(D/x)}$ is of the form

$$[\lambda y_1...y_m.C_{0(y_1/x_1,...,y_m/x_m)(D/x)}]$$

where for $1 \leqslant i \leqslant m$, either y_i is x_i, or y_i does not occur in C, and y_i is not free in D. Let C $v(\mathrm{D}/x)$-construct C and $C_{(D/x)}$ v-construct C′. We will show that C′ is C and so the Theorem is satisfied. Let C map $\langle \mathrm{X}_1, ..., \mathrm{X}_m \rangle$ to C_0. Then C_0 $v(\mathrm{D}/x)(\mathrm{X}_1/x_1, ..., \mathrm{X}_m/x_m)$-constructs C_0. By the 'Replacement of Free Variables' theorem, $C_{0(y_1/x_1,...,y_m/x_m)}$ $v(\mathrm{D}/x)(\mathrm{X}_1/y_1, ..., \mathrm{X}_m/y_m)$-constructs, and so also $v(\mathrm{X}_1/y_1, ..., \mathrm{X}_m/y_m)(\mathrm{D}/x)$-constructs C_0. Since $y_1, ..., y_m$ are not free in D, D $v(\mathrm{X}_1/y_1, ..., \mathrm{X}_m/y_m)$-constructs D. Hence (by the induction hypothesis),

$$C_{0(y_1/x_1,...,y_m/x_m)(D/x)}$$

$v(\mathrm{X}_1/y_1, ..., \mathrm{X}_m/y_m)$-constructs C_0, which means that C′ maps $\langle \mathrm{X}_1, ..., \mathrm{X}_m \rangle$ to C_0. The converse can be proven similarly.

In case (iii), C is of the form ${}^1C_0$, where C_0 is of rank $2 \leqslant r$. By the theorem '${}^1C \cong C$', ${}^1C_0$ is v-congruent with C_0 and it is thus easy to see that the Theorem is satisfied.

In case (iv), C is of the form ${}^2C_0$, where C_0 is of rank $2 \leqslant r$. Given (r_2), it is easy to see that the case with rank $r = 2$ satisfies the Theorem. With the few exceptions described below, even the cases with rank $2 < r$ are treated by (r_{+1}) as described above, and so the Theorem is satisfied.

To complete (r_{+1}), we must also cover exceptional constructions of the form (a) ${}^{20}C_0$, (b) ${}^{202}C_0$ and (c) ${}^{220}C_0$. If C's rank $r = 3$, the only possible form of such C is (a); if C's rank $4 \leqslant r$, C is of the form (a), (b), or (c). Now recall that ${}^{20}C_0$ is v-congruent with C_0 by the theorem '${}^{20}C \cong C$'. Without employing the theorem, we would wrongly deal with ${}^{20}C_0$ as a construction trivially satisfying the Theorem since ${}^0C_0$ (of rank r) has no free variable, cf. (1); but C_0 may contain a free variable which is thus free in ${}^{20}C_0$. In case (a), i.e. ${}^{20}C_0$, it is then easy to see that the Theorem is obviously satisfied, since, by the induction hypothesis, C_0 is of rank $r' < r$, for which the Theorem is already satisfied. Similarly for case (b), where ${}^{202}C_0$ is v-congruent with ${}^2C_0$. Moreover, the Theorem is satisfied even in case (c), since ${}^{220}C_0$'s subconstruction ${}^{20}C_0$ v-congruent with C_0 of rank $r' < r$, for which the Theorem is already satisfied.

Bibliography

[1] Martin Abadi et al. "Explicit Substitutions". In: *Journal of Functional Programming* 1.4 (1991), pp. 375–416. DOI: https://doi.org/10.1017/S0956796800000186.

[2] Theodora Achourioti et al., eds. *Unifying the Philosophy of Truth.* Springer, 2015. ISBN: 978-94-017-9672-9. DOI: https://doi.org/10.1007/978-94-017-9673-6.

[3] Natacha Alechina, Mark Jago, and Brian Logan. "Preference-Based Belief Revision for Rule-Based Agents". In: *Synthese* 165.1 (2008), pp. 159–177. DOI: https://doi.org/10.1007/s11229-008-9364-0.

[4] Joseph Almog. "Naming without Necessity". In: *The Journal of Philosophy* 83.4 (1986), pp. 210–242. DOI: https://doi.org/10.2307/2026532.

[5] Maria Aloni and Paul Dekker, eds. *The Cambridge Handbook of Formal Semantics (Cambridge Handbooks in Language and Linguistics).* Cambridge University Press, 2016. ISBN: 978-1107028395.

[6] C. Anthony Anderson. "General Intensional Logic". In: *Handbook of Philosophical Logic. Volume II: Extensions of Classical Logic.* Ed. by Dov M. Gabbay and Franz Guenthner. Springer, 1984, pp. 355–385. ISBN: 978-94-009-6261-3. DOI: https://doi.org/10.1007/978-94-009-6259-0_7.

[7] C. Anthony Anderson. "The Lesson of Kaplan's Paradox about Possible World Semantics". In: *The Philosophy of David Kaplan.* Ed. by Joseph Almog and Paolo Leonardi. Oxford University Press, 2009, pp. 355–385. ISBN: 978-0195367881. DOI: https://doi.org/10.1093/acprof:oso/9780195367881.003.0006.

[8] C. Anthony Anderson. "The Paradox of Knower". In: *Journal of Philosophy* 80.6 (1983), pp. 338–356. DOI: https://doi.org/10.2307/2026335.

[9] C. Anthony Anderson and Joseph Owens, eds. *Propositional Attitudes. The Role of Content in Logic, Language, and Mind.* Center for the Study of Language and Information, 1990. ISBN: 978-0937073506.

[10] Peter B. Andrews. *An Introduction to Mathematical Logic and Type Theory: To Truth Through Proof.* 2nd. Springer, 2002. ISBN: 978-1402007637. DOI: https://doi.org/10.1007/978-94-015-9934-4.

[11] Carlos Areces et al. "Completeness in Hybrid Type Theory". In: *Journal of Philosophical Logic* 43.2–3 (2014), pp. 209–238. DOI: https://doi.org/10.1007/s10992-012-9260-4.

[12] Sergei Artemov and Melvin Fitting. "Justification Logic". In: *The Stanford Encyclopedia of Philosophy.* Ed. by Edward N. Zalta. Winter 2016. 2018. URL: https://plato.stanford.edu/archives/win2016/entries/logic-justification.

[13] Sergei Artemov and Roman Kuznets. "Logical Omniscience as a Computational Complexity Problem". In: *Proceedings of the 12th Conference on Theoretical Aspects of Rationality and Knowledge.* ACM, 2009, pp. 14–23. ISBN: 978-1-60558-560-4. DOI: https://doi.org/10.1145/1562814.1562821.

[14] Sergei Artemov and Roman Kuznets. "Logical Omniscience as Infeasibility". In: *Annals of Pure and Applied Logic* 165.1 (2014), pp. 6–25. DOI: https://doi.org/10.1016/j.apal.2013.07.003.

[15] Sergei Artemov and Elena Nogina. "Introducing Justification into Epistemic Logic". In: *Journal of Logic and Computation* 15.6 (2005), pp. 1059–1073. DOI: https://doi.org/10.1093/logcom/exi053.

[16] Nicholas Asher. "Intentional Paradoxes and an Inductive Theory of Propositional Quantification". In: *Proceeding: TARK '90 Proceedings of the 3rd Conference on Theoretical aspects of Reasoning about Knowledge*. Ed. by Rorith Parikh. Morgan Kaufmann Publishers, 1990, pp. 11–28. ISBN: 1-55880-105-8.

[17] Nicholas Asher. "Meanings Don't Grow on Trees". In: *Journal of Semantics* 3.3 (1984), pp. 229–247. DOI: https://doi.org/10.1093/jos/3.3.229.

[18] Nicholas Asher and Hans Kamp. *The Knower's Paradox and the Logic of Attitudes*. Tech. rep. 1985.

[19] Andrew Bacon and Gabriel Uzquiano. "Some Results on the Limits of Thoughts". In: *Journal of Philosophical Logic* 47.6 (2018), pp. 991–999. DOI: https://doi.org/10.1007/s10992-018-9458-1.

[20] Andrew Jonathan Bacon, John Hawthorne, and Gabriel Uzquiano. "Higher-order Free Logic and the Prior-Kaplan Paradox". In: *Canadian Journal of Philosophy* 46.4–5 (2016), pp. 493–541. DOI: https://doi.org/10.1080/00455091.2016.1201387.

[21] Alexandru Baltag and Bryan Renne. "Dynamic Epistemic Logic". In: *The Stanford Encyclopedia of Philosophy*. Ed. by Edward N. Zalta. Winter 2016. 2016. URL: https://plato.stanford.edu/archives/win2016/entries/dynamic-epistemic.

[22] Alexandru Baltag, Bryan Renne, and Sonja Smets. "The Logic of Justified Belief, Explicit Knowledge, and Conclusive Evidence". In: *Annals of Pure and Applied Logic* 165.1 (2014), pp. 49–81. DOI: https://doi.org/10.1016/j.apal.2013.07.005.

[23] Hendrik P. Barendregt. *The Lambda Calculus: Its Syntax and Semantics*. North-Holland Publishing Company, 1992. ISBN: 978-0444875082.

[24] Hendrik P. Barendregt, Will Dekker, and Richard Statman. *Lambda Calculus with Types*. Cambridge University Press, 2013. ISBN: 978-0521766142.

[25] Jon Barwise. "On Branching Quantifiers in English". In: *Journal of Philosophical Logic* 8.1 (1979), pp. 47–80. DOI: https://doi.org/10.1007/BF00258419.

[26] Jon Barwise and Robin Cooper. "Generalized Quantifiers and Natural Language". In: *Linguistics and Philosophy* 4.2 (1981), pp. 159–219. DOI: https://doi.org/10.1007/978-94-009-2727-8_10.

[27] Jon Barwise and John Perry. *Situations and Attitudes*. MIT Press, 1983. ISBN: 978-0262021890.

[28] Rainer Bäuerle and Max J. Cresswell. "Propositional Attitudes". In: *Handbook of Philosophical Logic, vol. 10*. Ed. by Dov M. Gabbay and Franz Guenthner. 2nd. Kluwer Academic Publishers, 2003, pp. 121–141. ISBN: 978-90-481-6431-8. DOI: https://doi.org/10.1007/978-94-017-4524-6_4.

[29] George Bealer. "Fine-grained Type-free Intensionality". In: *Properties, Types and Meaning Vol. I*. Ed. by Gennaro Chierchia, Barbara H. Partee, and Raymond Turner. Kluwer, 1989, pp. 177–230. ISBN: 1-55608-067-0.

[30] George Bealer. *Quality and Concept*. Clarendon Press, 1982. ISBN: 978-0198244288.

[31] George Bealer and Uwe Mönnich. "Property Theories". In: *Handbook of Philosophical Logic, vol. 4*. Ed. by Dov Gabbay and Franz Guenthner. Kluwer Academic Publishers, 1989, pp. 133–251. ISBN: 978-94-010-7021-8. DOI: https://doi.org/10.1007/978-94-009-1171-0_2.

[32] JC Beall, ed. *Revenge of the Liar: New Essays on the Paradox*. Oxford University Press, 2008. ISBN: 978-0199233915.

[33] JC Beall, Michael Glanzberg, and David Ripley. "Liar Paradox". In: *The Stanford Encyclopedia of Philosophy*. Ed. by Edward N. Zalta. Fall 2017 Edition. 2017. URL: https://plato.stanford.edu/archives/fall2017/entries/liar-paradox.

[34] Hourya Benis-Sinaceur, Marco Panza, and Gabriel Sandu. "From Lagrange to Frege: Functions and Expressions". In: *Functions and Generality of Logic. Reflections on Dedekind's and Frege's Logicisms*. Ed. by Hourya Benis-Sinaceur, Marco Panza, and Gabriel Sandu. Springer, 2015, pp. 59–95. ISBN: 978-3-319-17108-1. DOI: https://doi.org/10.1007/978-3-319-17109-8_2.

[35] Johan van Benthem and Kees Doets. "Higher-Order Logic". In: *Handbook of Philosophical Logic*. Ed. by Dov M. Gabbay and Franz Guenthner. Springer, 2001, pp. 189–243. ISBN: 978-90-481-5717-4. DOI: https://doi.org/10.1007/978-94-015-9833-0_3.

[36] Johann F.A.K van Benthem and Alice G.B. ter Meulen, eds. *Handbook of Logic and Language*. 2nd edition. Elsevier, The MIT Press, 2011. ISBN: 978-0444537263.

[37] Johann F.A.K van Benthem and Eric Pacuit. "Dynamic Logics of Evidence-Based Beliefs". In: *Philosophy and Phenomenological Research* 99.1–3 (2011), pp. 61–92. DOI: https://doi.org/10.1007/s11225-011-9347-x.

[38] Christoph Benzmüller and Peter S. Andrews. "Church's Type Theory". In: *The Stanford Encyclopedia of Philosophy*. Ed. by Edward N. Zalta. Summer 2019. 2019. URL: https://plato.stanford.edu/archives/sum2019/entries/type-theory-church.

[39] Francesco Berto. *How to Sell a Contradiction. The Logic and Metaphysics of Inconsistency*. College Publications, 2007. ISBN: 978-1904987437.

[40] Francesco Berto. "Impossible Worlds and Propositions: Against the Parity Thesis". In: *The Philosophical Quarterly* 60.240 (2010), pp. 471–486. DOI: https://doi.org/10.1111/j.1467-9213.2009.627.x.

[41] Marta Bílková, Ondrej Majer, and Michal Peliš. "Epistemic Logics for Sceptical Agents". In: *Journal of Logic and Computation* 26.6 (2016), pp. 1815–1841. DOI: https://doi.org/10.1093/logcom/exv009.

[42] Marta Bílková et al. "The Logic of Resources and Capabilities". In: *The Review of Symbolic Logic* 11.2 (2018), pp. 371–410. DOI: https://doi.org/10.1017/S175502031700034X.

[43] Jens Christian Bjerring. "Impossible Worlds and Logical Omniscience: an Impossibility Result". In: *Synthese* 190.13 (2013), 2505–2524. DOI: https://doi.org/10.1007/s11229-011-0038-yS.

[44] Stephen Blamey. "Partial Logic". In: *Handbook of Philosophical Logic*. Ed. by Dov Gabbay and Franz Guenthner. Vol. 5. Springer, 2002, pp. 261–353. ISBN: 978-90-481-5927-7. DOI: https://doi.org/10.1007/978-94-017-0458-8_5.

[45] Dmitrii Anatoljevich Bochvar and Merrie Bergmann. "On a Three-valued Calculus and Its Applications to the Analysis of the Paradoxes of the Classical Extended Functional Calculus". In: *History and Philosophy of Logic* 2.1–2 (1981), pp. 87–112. DOI: https://doi.org/10.1080/01445348108837023.

[46] Adam Brandenburger and H. Jerome Keisler. "An Impossibility Theorem on Beliefs in Games". In: *Studia Logica* 84.2 (2006), pp. 211–240. DOI: `https://doi.org/10.1007/s11225-006-9011-z`.

[47] Robert Brandom. *Making It Explicit. Reasoning, Representing, and Discursive Commitment.* Harvard University Press, 1994. ISBN: 978-0674543300.

[48] Also Bressan. *A General Interpreted Modal Calculus.* Yale University Press, 1972. ISBN: 978-0300014297.

[49] Berit Brogaard and Joe Salerno. "Fitch's Paradox of Knowability". In: *The Stanford Encyclopedia of Philosophy.* Ed. by Edward N. Zalta. Winter 2013. 2013. URL: `https://plato.stanford.edu/archives/win2013/entries/fitch-paradox`.

[50] L.E.J. Brouwer. *Collected Works, Vol. I.* North-Holland, 1975. ISBN: 978-0-444-10643-8.

[51] L.E.J. Brouwer. *Collected Works, Vol. II.* North-Holland, 1976. ISBN: 978-0-7204-2076-0.

[52] Tyler Burge. "Belief *de re*". In: *Journal of Philosophy* 74.6 (1977), pp. 338–362. DOI: `https://doi.org/10.2307/2025871`.

[53] Tyler Burge. "Semantical Paradox". In: *Journal of Philosophy* 76.4 (1979), pp. 169–198. DOI: `https://doi.org/10.2307/2025724`.

[54] Libor Běhounek and Martina Daňková. "Towards Fuzzy Partial Set Theory". In: *16th International Conference, IPMU 2016, Eindhoven, The Netherlands, June 20 - 24, 2016, Proceedings, Part II.* Ed. by Joao Paulo Carvalho et al. Communications in Computer and Information Science. Springer, 2016, pp. 482–494. ISBN: 978-3-319-40580-3. DOI: `https://doi.org/10.1007/978-3-319-40580-3`.

[55] Andrea Cantini and Riccardo Bruni. "Paradoxes and Contemporary Logic". In: *The Stanford Encyclopedia of Philosophy.* Ed. by Edward N. Zalta. Fall 2017 Edition. 2018. URL: `https://plato.stanford.edu/archives/fall2017/entries/paradoxes-contemporary-logic`.

[56] Felice Cardone and J. Roger Hindley. "History of Lambda-calculus and Combinatory Logic". In: *Handbook of the History of Logic.* Ed. by Dov M. Gabbay and John Woods. Vol. 5. North Holland, 2009, pp. 723–817. ISBN: 978-0444516206.

[57] Rudolf Carnap. *Meaning and Necessity.* Second Edition. The University of Chicago Press, 1958.

[58] Walter Carnielli, Marcelo E. Coniglio, and Joao Marcos. "Logics of Formal Inconsistency". In: *Handbook of Philosophical Logic.* Ed. by Dov Gabbay and Franz Guenthner. 2nd edition. Vol. 14. Springer, 2007, pp. 15–107. ISBN: 978-1-4020-6323-7.

[59] Massimiliano Carrara and Davide Fassio. "Why Knowledge Should Not Be Typed: An Argument against the Type Solution to the Knowability Paradox". In: *Theoria* 77.2 (2011), pp. 180–193. DOI: `https://doi.org/10.1111/j.1755-2567.2011.01100.x`.

[60] Hector-Neri Castañeda. "On the Logic of Attributions of Self-Knowledge to Others". In: *The Journal of Philosophy* 65.15 (1968), pp. 439–456. DOI: `https://doi.org/10.2307/2024296`.

[61] David Chalmers. "Propositions and Attitude Ascriptions: A Fregean Account". In: *Noûs* 45.4 (2011), pp. 595–639. DOI: `https://doi.org/10.1111/j.1468-0068.2010.00788.x`.

[62] Stergios Chatzikyriakidis and Zhaohui Luo, eds. *Modern Perspectives in Type-Theoretical Semantics.* Springer, 2017. ISBN: 978-3319843971.

[63] Gennaro Chierchia and Sally McConnell-Ginet. *Meaning and Grammar: An Introduction to Semantics.* 2nd. MIT Press, 2000. ISBN: 978-0262531641.

[64] Gennaro Chierchia and Raymond Turner. "Semantics and Property Theory". In: *Linguistic and Philosophy* 11.2–36 (1990), pp. 95–120. DOI: https://doi.org/10.1007/BF00632905.

[65] Charles Chihara. *Ontology and the Vicious-Circle Principle.* Cornell University Press, 1973. ISBN: 978-0801407277.

[66] Alonzo Church. "A Comparison of Russell's Resolution of the Semantical Antinomies with that of Tarski". In: *Journal of Symbolic Logic* 41.4 (1976), pp. 747–760. DOI: https://doi.org/10.2307/2272393.

[67] Alonzo Church. "A Formulation of the Logic of Sense and Denotation". In: *Structure, Method and Meaning (Essays in Honor of Henry M. Sheffer).* Ed. by Paul Henle, Horace Meyer Kallen, and Susanne Langer. Liberal Arts Press, 1951, pp. 3–34.

[68] Alonzo Church. "A Formulation of the Simple Theory of Types". In: *The Journal of Symbolic Logic* 5.2 (1940), pp. 56–68. DOI: https://doi.org/10.2307/2266170.

[69] Alonzo Church. "A Note on the Entscheidungsproblem". In: *Journal of Symbolic Logic* 1.1 (1936), pp. 40–41. DOI: https://doi.org/10.2307/2269326.

[70] Alonzo Church. "A Set of Postulates for the Foundation of Logic". In: *The Annals of Mathematics* 33.2 (1932), pp. 346–366. DOI: https://doi.org/10.2307/1968337.

[71] Alonzo Church. "An Unsolvable Problem of Elementary Number Theory". In: *American Journal of Mathematics* 58.2 (1936), pp. 345–364. DOI: https://doi.org/10.2307/2268571.

[72] Alonzo Church. *Introduction to Mathematical Logic.* Princeton University Press, 1956. ISBN: 978-0691029061.

[73] Alonzo Church. "On Carnap's Analysis of Statements of Assertion and Belief". In: *Analysis* 10.5 (1950), pp. 97–99. DOI: https://doi.org/10.2307/3326684.

[74] Alonzo Church. "Referee Reports on Fitch's "A Definition of Value"". In: *New Essays on the Knowability Paradox.* Ed. by Joe Salerno. Oxford University Press, 2009, pp. 13–20. ISBN: 978-0199285495. DOI: https://doi.org/10.1093/acprof:oso/9780199285495.003.0002.

[75] Alonzo Church. "Russell's Theory of Identity of Propositions". In: *Philosophia Naturalis* 21.2/4 (1984), pp. 513–522.

[76] Alonzo Church. "Schröder's Anticipation of the Simple Theory of Types". In: *Erkenntnis* 10.3 (1976), pp. 407–411. DOI: https://doi.org/10.1007/BF00214734.

[77] Alonzo Church. *The Calculi of Lambda-Conversion.* Princeton University Press, 1985. ISBN: 0-691-08394-0.

[78] Pavel Cmorej. *Na pomedzí logiky a filozofie [On the Borderline of Logic and Philosophy].* Veda, 2001. ISBN: 80-224-0699-6.

[79] Irving M. Copi. *The Theory of Logical Types.* Routledge, 1971. ISBN: 978-0710070265.

[80] Thierry Coquand. "Type Theory". In: *The Stanford Encyclopedia of Philosophy.* Ed. by Edward N. Zalta. Summer 2015. 2015. URL: https://plato.stanford.edu/archives/sum2015/entries/type-theory.

[81] Thierry Coquand and Gérard Huet. "The Calculus of Constructions". In: *Information and Computation* 36.2–3 (1988), pp. 95–120. DOI: https://doi.org/10.1016/0890-5401(88)90005-3.

[82] Sean Crawford. "Quantifiers and Propositional Attitudes: Quine Revisited". In: *Synthese* 160 (2008), pp. 75–96. DOI: https://doi.org/10.1007/s11229-006-9080-6.

[83] Max J. Cresswell. "Hyperintensional Logic". In: *Studia Logica* 34.1 (1975), pp. 26–38. DOI: https://doi.org/10.1007/BF02314421.

[84] Max J. Cresswell. "Quotational Theories of Propositional Attitudes". In: *Journal of Philosophical Logic* 9.1 (1980), pp. 17–40. DOI: https://doi.org/https://doi.org/10.1007/978-94-015-7778-6_6.

[85] Max J. Cresswell. *Structured Meanings*. MIT Press (A Bradford Book), 1985. ISBN: 978-0262518956.

[86] Max J. Cresswell and Arnim von Stechow. "*De re* Belief Generalized". In: *Studia Logica* 5.4 (1982), pp. 503–535. DOI: https://doi.org/10.1007/BF00355585.

[87] Charles B. Cross. "The Paradox of the Knower without Epistemic Closure". In: *Mind* 110.438 (2001), pp. 319–333. DOI: https://doi.org/10.1093/mind/110.438.319.

[88] Charles B. Cross. "The Paradox of the Knower without Epistemic Closure – Corrected". In: *Mind* 121.482 (2012), pp. 457–466. DOI: https://doi.org/10.1093/mind/fzs067.

[89] Haskell B. Curry. "Grundlagen der Kombinatorischen Logik". In: *American Journal of Mathematics* 52.3 (1930), pp. 789–834. DOI: https://doi.org/10.2307/2370619.

[90] Haskell B. Curry and Robert Feys. *Combinatory Logic*. Vol. I. North Holland, 1958. DOI: https://doi.org/10.1017/S0022481200114203.

[91] Dirk Van Dalen. "Intuitionistic Logic". In: *The Blackwell Guide to Philosophical Logic*. Ed. by Lou Goble. Blackwell, 2001, pp. 224–257. ISBN: 978-0631206934.

[92] Martin Davies. *Meaning, Quantification and Necessity: Themes in Philosophical Logic*. Routledge and Kegan Paul, 1981. ISBN: 978-0710007599.

[93] Walter Dean and Hidenori Kurokawa. "The Paradox of the Knower Revisited". In: *Annals of Pure and Applied Logic* 165.1 (2014), pp. 199–224. DOI: https://doi.org/10.1016/j.apal.2013.07.010.

[94] Hans van Ditmarsch, Wiebe van der Hoek, and Barteld Kooi. *Dynamic Epistemic Logic*. Synthese Library: Studies in Epistemology, Logic, Methodology, and Philosophy of Science. Springer, 2008. ISBN: 978-1-4020-5838-7. DOI: https://doi.org/10.1007/978-1-4020-5839-4.

[95] Hans van Ditmarsch et al., eds. *Handbook of Epistemic Logic*. College Publications, 2015. ISBN: 978-1848901582.

[96] Kosta Došen. "Identity of Proofs Based on Normalization and Generality". In: *The Bulletin of Symbolic Logic* 9.4 (2003), pp. 477–503. DOI: https://doi.org/10.2178/bsl/1067620091.

[97] David R. Dowty, Robert E. Wall, and Stanley Peters. *Introduction to Montague Semantics*. Kluwer, 1992. ISBN: 978-90-277-1142-7. DOI: https://doi.org/10.1007/978-94-009-9065-4.

[98] Theodore Drange. "The Paradox of the Non-Communicator". In: *Philosophical Studies* 15.6 (1964), pp. 92–96. DOI: https://doi.org/10.1007/BF00420415.

[99] Ho Ngoc Duc. "Reasoning about Rational, but not Logically Omniscient, Agents". In: *Journal of Logic and Computation* 7.5 (1997), pp. 633–648. DOI: https://doi.org/10.1093/logcom/7.5.633.

[100] Ho Ngoc Duc. "Resource-Bounded Reasoning about Knowledge". Ph.D. thesis. 2001.

[101] Michael Dummett. *The Logical Basis of Metaphysics*. Harward University Press, 1993. ISBN: 978-0674537866.

[102] Marie Duží. "Deduction in TIL: From Simple to Ramified Hierarchy of Types". In: *Organon F* Supplementary Issue 2 (2013), pp. 5–36.

[103] Marie Duží. "Do We Have to Deal with Partiality?" In: *Miscellanea Logica* V (2003), pp. 45–76.

[104] Marie Duží. "If Structured Propositions Are Logical Procedures Then How Are Procedures Individuated?" In: *Synthese* 196.4 (2019), pp. 1249–1283. DOI: https://doi.org/10.1007/s11229-017-1595-5.

[105] Marie Duží. "Presuppositions and Two Kinds of Negation". In: *Logique et Analyse* 239 (2017), pp. 245–226. DOI: https://doi.org/10.2143/LEA.239.0.3237153.

[106] Marie Duží and Bjørn Jespersen. "Procedural Isomorphism, Analytic Information and β-Conversion by Value". In: *Logic Journal of IGPL* 21.2 (2013), pp. 291–308. DOI: https://doi.org/10.1093/jigpal/jzs044.

[107] Marie Duží and Bjørn Jespersen. "Transparent Quantification into Hyperintensional Objectual Attitudes". In: *Synthese* 192.3 (2015), pp. 635–677. DOI: https://doi.org/10.1007/s11229-014-0578-z.

[108] Marie Duží and Bjørn Jespersen. "Transparent Quantification into Hyperpropositional Contexts de re". In: *Logique et Analyse* 55.220 (2012), pp. 291–308.

[109] Marie Duží, Bjørn Jespersen, and Jaroslav Müller. "Epistemic Closure and Inferable Knowledge". In: *The Logica Yearbook 2004*. Ed. by Libor Běhounek and Marta Bílková. Filozofie, 2005, pp. 125–140. ISBN: 80-7007-208-3.

[110] Marie Duží, Bjørn Jespersen, and Pavel Materna. *Procedural Semantics for Hyperintensional Logic: Foundations and Applications of Transparent Intensional Logic*. Springer, 2010. ISBN: 978-90-481-8811-6. DOI: https://doi.org/10.1007/978-90-481-8812-3.

[111] Rolf A. Eberle. "A Logic of Believing, Knowing, and Inferring". In: *Synthese* 26.3 (1974), pp. 356–382. DOI: https://doi.org/10.1007/BF00883100.

[112] Walter Edelberg. "Propositions, Circumstance, and Objects". In: *Journal of Philosophical Logic* 23.1 (2010), pp. 1–34. DOI: https://doi.org/10.1007/BF01417956.

[113] Paul Égré. "Epistemic Logic". In: *The Bloomsbury Companion to Philosophical Logic*. Ed. by Leo Horsten and Richard Pettigrew. second edition. 2014, pp. 503–542. ISBN: 978-1472523020.

[114] Paul Égré. "The Knower Paradox in the Light of Provability Interpretations of Modal Logic". In: *Journal of Logic, Language and Information* 14.1 (2005), pp. 13–48. DOI: https://doi.org/10.1007/s10849-004-6406-y.

[115] Matti Eklund. "Regress, Unity, Facts, and Propositions". In: *Synthese* 196.4 (2019), pp. 1225–1247. DOI: https://doi.org/10.1007/s11229-016-1155-4.

[116] Paul Elbourne. "Why Propositions Might be Sets of Truth-supporting Circumstances". In: *Journal of Philosophical Logic* 39.1 (2010), pp. 101–111. DOI: https://doi.org/10.1007/s10992-009-9112-z.

[117] Jennifer J. Elgot-Drapkin, Michael Miller, and Donald Perlis. "Memory, Reason, and Time: the Step-Logic Approach". In: *Philosophy and AI: Essays at the Interface*. Ed. by Cummings and Pollock. 1991, pp. 79–103.

[118] Jennifer J. Elgot-Drapkin and Donald Perlis. "Reasoning Situated in Time I: Basic Concepts". In: *Journal of Experimental and Theoretical Artificial Intelligence* 2.1 (2010), pp. 75–98. DOI: https://doi.org/10.1080/09528139008953715.

[119] Ronald Fagin and Joseph Y. Halpern. "Belief, Awareness, and Limited Reasoning". In: *Artificial Intelligence* 34.1 (1988), pp. 39–76. DOI: https://doi.org/10.1016/0004-3702(87)90003-8.

[120] Ronald Fagin et al. *Reasoning about Knowledge*. MIT Press, 1995. ISBN: 978-0262061629.

[121] William M. Farmer. "A Partial Functions Version of Church's Simple Theory of Types". In: *Journal of Symbolic Logic* 55.3 (1990), pp. 1269–1291. DOI: https://doi.org/10.2307/2274487.

[122] Federico L. G. Faroldi. "Co-hyperintensionality". In: *Hyperintensionality and Normativity*. Springer, 2019, pp. 61–75. DOI: https://doi.org/10.1007/978-3-030-03487-0_3.

[123] Solomon Feferman. "Definedness". In: *Erkenntnis* 43.3 (1995), pp. 295–320. DOI: https://doi.org/10.1007/BF01135376.

[124] Solomon Feferman. "Predicativity". In: *The Oxford Handbook of Philosophy of Mathematics and Logic*. Ed. by Stewart Shapiro. Oxford University Press, 2005, pp. 590–624.

[125] Frederic B. Fitch. "A Logical Analysis of Some Value Concepts". In: *The Journal of Symbolic Logic* 28.2 (1963), pp. 135–142. DOI: https://doi.org/10.2307/2271594.

[126] Frederic B. Fitch. "Self-reference in Philosophy". In: *Mind* 55.217 (1964), pp. 64–73. DOI: https://doi.org/10.1093/mind/LV.219.64.

[127] Melvin Fitting. "First-Order Intensional Logic". In: *Annals of Pure and Applied Logic* 127.1–3 (2004), pp. 171–193. DOI: https://doi.org/10.1016/j.apal.2003.11.014.

[128] Melvin Fitting. "Intensional Logic". In: *The Stanford Encyclopedia of Philosophy*. Ed. by Edward N. Zalta. Summer 2015. 2015. URL: http://plato.stanford.edu/archives/sum2015/entries/logic-intensional.

[129] Melvin Fitting and Richard L. Mendelsohn. *First-Order Modal Logic*. Kluwer, 1998. ISBN: 978-0792353355.

[130] Peter Fletcher. *Truth, Proof and Infinity. A Theory of Constructions and Constructive Reasoning*. Kluwer, 1998. ISBN: 978-0792352624.

[131] Salvatore Florio and Julien Murzi. "The Paradox of Idealization". In: *Analysis* 69.3 (2009), pp. 461–469. DOI: https://doi.org/10.1093/analys/anp069.

[132] Graeme Forbes. *Attitude Problems: An Essay on Linguistic Intentionality*. Oxford University Press, 2006. ISBN: 978-0199274949.

[133] Graeme Forbes. "The Indispensability of Sinn". In: *The Philosophical Review* 99.4 (2009), pp. 535–564. DOI: https://doi.org/10.2307/2185616.

[134] Chris Fox and Shalom Lappin. *Foundations of Intensional Semantics*. Blackwell, 2005. ISBN: 978-0631233763.

[135] Nissim Francez. *Proof-theoretic Semantics*. College Publications, 2015. ISBN: 978-1848901834.

[136] Nissim Francez and Roy Dyckhoff. "Proof-theoretic Semantics for a Natural Language Fragment". In: *Linguistics and Philosophy* 33.6 (2010), pp. 447–477. DOI: https://doi.org/10.1007/s10988-011-9088-3.

[137] Gottlob Frege. *Begriffsschrift, eine der Arithmetischen Nachgebildete Formalsprache des Reinen Denken*. Verlag von Louis Nebert, 1879.

[138] Gottlob Frege. "Der Gedanke. Eine logische Untersuchung". In: *Beiträge zur Philosophie des deutschen Idealismus* 2 (1918–1919), pp. 58–77.

[139] Gottlob Frege. *Funktion und Begriff.* H. Pohle, 1891.

[140] Gottlob Frege. *Grundlagen der Arithmetik. Eine logisch-mathematische Untersuchung über den Begriff der Zahl.* Verlag von Wilhelm Koebner, 1879.

[141] Gottlob Frege. *Philosophical and Mathematical Correspondence.* Ed. by Gottfried Gabriel et al. Blackwell, 1980. ISBN: 978-0226261973.

[142] Hans Hermes, Friedrich Kambartel, and Friedrich Kaulbach, eds. *Posthumous Writings.* University of Chicago Press, 1979. ISBN: 978-0226261997.

[143] Gottlob Frege. "Über Sinn und Bedeutung". In: *Zeitschrift für Philosophie und philosophishe Kritik* 100 (1892), pp. 25–50.

[144] Peter Fritz. "Higher-Order Contingetism, Part 3: Expressive Limitation". In: *Journal of Philosophical Logic* 47.4 (2018), pp. 649–671. DOI: https://doi.org/10.1007/s10992-017-9443-0.

[145] Daniel Gallin. *Intensional and Higher-Order Modal Logic.* North Holland, 1975. ISBN: 978-1483274737.

[146] L.T.F. Gamut. *Logic, Language, and Meaning. Volume 2 Intensional Logic and Logical Grammar.* The University of Chicago Press, 1991. ISBN: 978-0226280882.

[147] Manuel García-Carpintero and Bjørn Jespersen. "Introduction: Primitivism versus Reductionism about the Problem of the Unity of the Proposition". In: ed. by Manuel García-Carpintero and Bjørn Jespersen. Vol. 196. 2019, pp. 1209–1224. DOI: https://doi.org/10.1007/s11229-018-1727-6.

[148] Peter Gärdenfors. *Knowledge in Flux. Modeling the Dynamics of Epistemic States.* MIT Press, 1988. ISBN: 978-0262071093.

[149] Peter Gärdenfors and Hans Rott. "Belief Revision". In: *Handbook of Logic in Artificial Intelligence and Logic Programming, Vol. IV: Epistemic and Temporal Reasoning.* Ed. by Dov Gabbay, C.J. Hogger, and J.A. Robinson. Oxford University Press, 1995, pp. 35–132. ISBN: 0-19-853791-3.

[150] Richard Gaskin. *The Unity of the Proposition.* Oxford University Press, 2008. ISBN: 978-0199239450.

[151] Peter T. Geach. "Intentional Identity". In: *The Journal of Philosophy* 64.20 (1967), pp. 627–632. DOI: https://doi.org/10.2307/2024459.

[152] Margaret E. Szabo, ed. *The Collected Papers of Gerhard Gentzen.* North-Holland Pub. Co., 1969. ISBN: 978-0-444-53419-4.

[153] Gerhard Gentzen. "Untersuchungen über das logische Schließen". In: *Mathematische Zeitschrift* 39 (1934–1935), pp. 176–210,405–431.

[154] Pierdaniele Giaretta. "Liar, Reducibility and Language". In: *Synthese* 117.3 (1998), pp. 355–374. DOI: https://doi.org/10.1023/A:1005076008996.

[155] Pierdaniele Giaretta. "The Paradox of Knowability from a Russellian Perspective". In: *Prolegomena* 8.2 (2009), pp. 141–158.

[156] Jean-Yves Girard, Paul Taylor, and Yves Lafont. *Proofs and Types.* Cambridge University Press, 1989. ISBN: 978-0521371810.

[157] Kurt Gödel. "Russell's Mathematical Logic". In: *The Philosophy of Bertrand Russell.* Ed. by Paul Arthur Schilpp. Northwestern University, 1944, pp. 125–153.

[158] Warren Goldfarb. "Russell's Reasons for Ramification". In: *Rereading Russell: Essays in Russell's Metaphysics and Epistemology.* Ed. by C. Wade Savage and C. Anthony Anderson. University of Minnesota Press, 1989, pp. 22–40. ISBN: 978-0816669332.

[159] Patrick Grim. *The Incomplete Universe.* MIT Press, 1991. ISBN: 0-262-07134-7.

[160] Patrick Grim. "Truth, Omniscience, and the Knower". In: *Philosopical Studies* 54.1 (1988), pp. 9–41. DOI: https://doi.org/10.1007/BF00354176.

[161] Anil Gupta and Nuel Belnap. *The Revision Theory of Truth.* MIT Press, 1993. ISBN: 978-0262071444.

[162] Volker Halbach. "On a Side Effect of Solving Fitch's Paradox by Typing Knowledge". In: *Analysis* 68.2 (2008), pp. 114–120. DOI: https://doi.org/10.1093/analys/68.2.114.

[163] Volker Halbach, Hannes Leitgeb, and Philip Welch. "Possible Worlds Semantics for Modal Notions Conceived as Predicates". In: *Journal of Philosophical Logic* 32.2 (2003), pp. 179–223. DOI: https://doi.org/10.1023/A:1023080715357.

[164] Volker Halbach and Philip Welch. "Necessities and Necessary Truths: A Prolegomenon to the Use of Modal Logic in the Analysis of Intensional Notions". In: *Mind* 118.469 (2008), pp. 71–100. DOI: https://doi.org/10.1093/mind/fzn030.

[165] Joseph Y. Halpern, Yoram Moses, and Moshe Y. Vardi. "Algorithmic Knowledge". In: *Proceeding TARK '94 Proceedings of the 5th Conference on Theoretical Aspects of Reasoning About Knowledge.* Ed. by Ronald Fagin. 1994, pp. 255–266. ISBN: 978-1483214535. DOI: https://doi.org/10.1016/B978-1-4832-1453-5.50022-2.

[166] Joseph Y. Halpern and Riccardo Pucella. "Dealing with Logical Omniscience: Expressiveness and Pragmatics". In: *Artificial Intelligence* 175.1 (2011), pp. 220–235. DOI: https://doi.org/10.1016/j.artint.2010.04.009.

[167] W.D. Hart. "Invincible Ignorance". In: *New Essays on the Knowability Paradox.* Ed. by Joe Salerno. Oxford University Press, 2009, pp. 321–323. ISBN: 978-0199285495. DOI: 10.1093/acprof:oso/9780199285495.003.0020.

[168] Allen Hazen. "Predicative Logics". In: *Handbook of Philosophical Logic.* Ed. by Dov M. Gabbay and Franz Guenther. Kluwer, 1983, pp. 331–407. ISBN: 978-94-009-7068-7. DOI: https://doi.org/10.1007/978-94-009-7066-3_5.

[169] Vincent Hendricks and John Symons. "Epistemic Logic". In: *The Stanford Encyclopedia of Philosophy.* Ed. by Edward N. Zalta. Fall 2015. 2015. URL: http://plato.stanford.edu/entries/logic-epistemic.

[170] Vincent F. Hendricks and John Symons. "Where's the Bridge? Epistemology and Epistemic Logic". In: *Philosophical Studies* 128.1 (2006), pp. 137–167. DOI: https://doi.org/10.1007/s11098-005-4060-0.

[171] Leon Henkin. "Completeness in the Theory of Types". In: *The Journal of Symbolic Logic* 15.2 (1950), pp. 81–91. DOI: https://doi.org/10.2307/2266967.

[172] Klaus von Heusinger, Claudia Maienborn, and Paul Portner, eds. *Semantics: An International Handbook of Natural Language Meaning.* De Gruyter Mouton, 2011. ISBN: 978-3-11-018470-9.

[173] David Hilbert and Paul Bernays. *Grundlagen der Mathematik.* Springer Verlag, 1939. ISBN: 978-3-540-05110-7.

[174] James R. Hindley and Jonathan P. Seldin. *Lambda-Calculus and Combinators, an Introduction.* Cambridge University Press, 2008. ISBN: 978-0521898850.

[175] Jaakko Hintikka. "Impossible Possible Worlds Vindicated". In: *Journal of Philosophical Logic* 4.4 (1975), pp. 475–484. DOI: https://doi.org/10.1007/BF00558761.

[176] Jaakko Hintikka. "Individuals, Possible Worlds, and Epistemic Logic". In: *Noûs* 1.1 (1967), pp. 33–62. DOI: https://doi.org/10.2307/2214711.

[177] Jaakko Hintikka. *Knowledge and Belief: An Introduction to the Logic of the Two Notions.* Cornell University Press, 1962. ISBN: 1-904987-08-7.

[178] Jaakko Hintikka. "Semantics for Propositional Attitudes". In: *Model for Modalities*. D. Reidel Publishing Company, 1969, pp. 87–111. ISBN: 978-90-277-0598-3. DOI: https://doi.org/10.1007/978-94-010-1711-4_6.

[179] Jaakko Hintikka and Merrill B.P. Hintikka. "On Sense, Reference, and the Objects of Knowledge". In: *The Logic of Epistemology and the Epistemology of Logic. Selected Essays*. Springer, 1989, pp. 45–61. ISBN: 978-0-7923-0041-0. DOI: https://doi.org/10.1007/978-94-009-2647-9_4.

[180] Harold T. Hodes. "Why Ramify?" In: *Notre Dame Journal of Formal Logic* 56.2 (2015), pp. 379–415. DOI: https://doi.org/10.1215/00294527-2864352.

[181] William A. Howard. "The Formulae-As-Types Notion of Construction". In: *The Curry-Howard Isomorphism*. Ed. by Philippe De Groote. 1995, pp. 479–490.

[182] Andrzej Indrzejczak. "Natural Deduction". In: *The Internet Encyclopedia of Philosophy*. Ed. by James Fieser and Bradley Dowden. 2018. URL: https://www.iep.utm.edu/nat-ded.

[183] Andrzej Indrzejczak. *Natural Deduction, Hybrid Systems and Modal Logics*. Springer, 2010.

[184] A.D. Irvine. "Principia Mathematica". In: *The Stanford Encyclopedia of Philosophy*. Ed. by Edward N. Zalta. Winter 2016. 2016. URL: https://plato.stanford.edu/archives/win2016/entries/principia-mathematica.

[185] Bart Jacobs. *Categorical Logic and Type Theory*. Studies in Logic and the Foundations of Mathematics 141. North Holland, 1999. ISBN: 0-444-50170-3.

[186] Mark Jago. "Closure on Knowability". In: *Analysis* 70.4 (2010), pp. 648–659. DOI: https://doi.org/10.1093/analys/anq067.

[187] Mark Jago. "Hyperintensional Propositions". In: *Synthese* 192.3 (2015), pp. 585–601. DOI: https://doi.org/10.1007/s11229-014-0461-y.

[188] Mark Jago. *The Impossible: An Essay on Hyperintensionality*. Oxford University Press, 2014. ISBN: 978-0198709008. DOI: https://doi.org/10.1093/acprof:oso/9780198709008.001.0001.

[189] Theo M.V. Janssen. "Montague Semantics". In: *The Stanford Encyclopedia of Philosophy*. Ed. by Edward N. Zalta. Spring 2017. 2017. URL: https://plato.stanford.edu/archives/spr2017/entries/montague-semantics.

[190] Stanislaw Jaśkowski. "On the Rules of Suppositions in Formal Logic". In: *Studia Logica* 1.4 (1934), pp. 232–258.

[191] Bjørn Jespersen. "Anatomy of Propositions". In: *Synthese* 196 (4 2019), pp. 1285–1324. DOI: https://doi.org/10.1007/s11229-017-1512-y.

[192] Bjørn Jespersen. "How Hyper are Hyperpropositions?" In: *Language and Linguistics Compass* 4.2 (2014), pp. 96–106. DOI: https://doi.org/10.1111/j.1749-818X.2009.00181.x.

[193] Bjørn Jespersen. "Predication and Extensionalization". In: *Journal of Philosophical Logic* 37 (5 2008), pp. 479–499. DOI: https://doi.org/10.1007/s10992-007-9079-6.

[194] Bjørn Jespersen. "Recent Work on Structured Meaning and Propositional Unity". In: *Philosophy Compass* 7.9 (2012), pp. 620–630. DOI: https://doi.org/10.1111/j.1747-9991.2012.00509.x.

[195] Bjørn Jespersen. "Structured Lexical Concepts, Property Modifiers, and Transparent Intensional Logic". In: *Philosophical Studies* 172.2 (2015), pp. 321–345. DOI: https://doi.org/10.1007/s11098-014-0305-0.

[196] Bjørn Jespersen, Massimilliano Carrara, and Marie Duží. "Iterated Privation and Positive Modification". In: *Journal of Applied Logic* 25.Supplement (2017), S48–S71. DOI: https://doi.org/10.1016/j.jal.2017.12.004.

[197] Eric Johannesson. "Partial Semantics for Quantified Modal Logic". In: *Journal of Philosophical Logic* 47.6 (2018), pp. 1049–1060. DOI: https://doi.org/10.1007/s10992-018-9461-6.

[198] Darryl Jung. "Russell, Presupposition, and the Vicious-Circle Principle". In: *Notre Dame Journal of Formal Logic* 40.1 (1999), pp. 55–80. DOI: https://doi.org/10.1305/ndjfl/1039096305.

[199] Reinhard Kahle and Peter Schroeder-Heister, eds. *Proof-Theoretic Semantics.* Vol. 148. Synthese, 2006.

[200] Fairouz Kamareddine, Twan Laan, and Rob Nederpelt. "A History of Types". In: *Handbook of the History of Logic.* Ed. by Dov M. Gabbay, Francis J. Pelletier, and John Woods. Vol. 11. 2012, pp. 451–511. ISBN: 978-0-444-52937-4. DOI: https://doi.org/10.1016/B978-0-444-52937-4.50009-5.

[201] Fairouz Kamareddine, Twan Laan, and Rob Nederpelt. *A Modern Perspective on Type Theory. From Its Origins until Today.* Springer, 2004. ISBN: 978-1402023347.

[202] Fairouz Kamareddine, Twan Laan, and Rob Nederpelt. "Types in Logic and Mathematics Before 1940". In: *Bulletin of Symbolic Logic* 8.2 (2002), pp. 185–245. DOI: https://doi.org/10.2178/bsl/1182353871.

[203] David Kaplan. "A Problem in Possible Worlds Semantics". In: *Modality, Morality and Belief: Essays in Honor of Ruth Barcan Marcus.* Ed. by Walter Sinnott-Armstrong, Diana Raffman, and Nicholas Asher. Cambridge University Press, 1995, pp. 41–52. ISBN: 0-521-44082-3.

[204] David Kaplan. "How to Russell a Frege-Church". In: *The Journal of Philosophy* 72.19 (1975), pp. 716–729. DOI: https://doi.org/10.2307/2024635.

[205] David Kaplan. "Quantifying In". In: *Synthese* 19.1–2 (1968), pp. 178–214. DOI: https://doi.org/10.1007/BF00568057.

[206] David Kaplan and Richard Montague. "A Paradox Regained". In: *Notre Dame Journal of Formal Logic* 1.3 (1960), pp. 79–90. DOI: https://doi.org/10.1305/ndjfl/1093956549.

[207] Lorraine J. Keller. "What Propositional Structure Could Not Be". In: *Synthese* 196.4 (2019), pp. 1529–1553. DOI: https://doi.org/10.1007/s11229-017-1585-7.

[208] Jeffrey C. King. "On Fineness of Grain". In: *Philosopical Studies* 163.3 (2013), pp. 763–781. DOI: https://doi.org/10.1007/s11098-011-9844-9.

[209] Jeffrey C. King. "Structured Propositions". In: *The Stanford Encyclopedia of Philosophy.* Ed. by Edward N. Zalta. Winter 2016. 2016. URL: https://plato.stanford.edu/archives/win2016/entries/propositions-structured.

[210] Jeffrey C. King. "Structured Propositions and Sentence Structure". In: *Journal of Philosopical Logic* 25.5 (1996), pp. 495–521. DOI: https://doi.org/10.1007/BF00257383.

[211] Jeffrey C. King. *The Nature and Structure of Content.* Oxford University Press, 2007. ISBN: 978-0199226061.

[212] Stephen C. Kleene. *Introduction to Metamathematics.* North-Holland, 1952. ISBN: 0720421039.

[213] Stephen C. Kleene and Joseph B. Rosser. "The Inconsistency of Certain Formal Logics". In: *The Annals of Mathematics* 36.3 (1935), pp. 630–636. DOI: https://doi.org/10.2307/1968646.

[214] Kevin C. Klement. "Russell-Myhill Paradox". In: *The Internet Encyclopedia of Philosophy*. Ed. by James Fieser and Bradley Dowden. 2018. URL: http://www.iep.utm.edu/par-rusm.

[215] Petr Kolář and Vladimír Svoboda. "Ascribing An Action". In: *From the Logical Point of View* 1 (1992), pp. 34–60.

[216] Kurt Konolige. *A Deduction Model of Belief*. Morgan Kaufmann Publishers, 1986. ISBN: 0934613087.

[217] Kurt Konolige. "What Awareness Isn't: A Sentential View of Implicit and Explicit Belief". In: *Journal of Symbolic Logic* 53.2 (1986), pp. 241–250. DOI: https://doi.org/10.1016/B978-0-934613-04-0.50019-3.

[218] Robert C. Koons. *Paradoxes of Belief and Strategic Rationality*. Cambridge University Press, 1992. ISBN: 978-0521412698.

[219] Saul A. Kripke. *Naming and Necessity*. Basil Blackwell, 1980. ISBN: 0-674-59846-6.

[220] Saul A. Kripke. "Outline of a Theory of Truth". In: *The Journal of Philosophy* 72.19 (1975), pp. 690–716. DOI: https://doi.org/10.2307/2024634.

[221] Petr Kuchyňka and Jiří Raclavský. *Pojmy a vědecké teorie [Concepts and Scientific Theories]*. Masarykova univerzita, 2014. ISBN: 978-80-210-6791-2.

[222] Steven T. Kuhn. "Quantifiers as Modal Operators". In: *Studia Logica* 34.2–3 (1980), pp. 145–158. DOI: https://doi.org/10.1007/BF00370318.

[223] Wolfgang Künne, Albert Newen, and Martin Anduschus. *Direct Reference, Indexicality, and Propositional Attitudes*. Center for Study of Language and Information, 1997. ISBN: 978-1575860718.

[224] Twan Laan and Rob Nederpelt. "A Modern Elaboration of the Ramified Theory of Types". In: *Studia Logica* 57.2–3 (1996), pp. 243–278. DOI: https://doi.org/10.1007/BF00370835.

[225] Gregory Landini. *Russell's Hidden Substitutional Theory*. Oxford University Press, 1998. ISBN: 978-0195116830.

[226] Tore Langholm. "How Different is Partial Logic?" In: *Partiality, Modality, and Nonmonotonicity*. Ed. by Patrick Doherty. CSLI, 1996, pp. 3–43. ISBN: 978-1-57586-030-9.

[227] Serge Lappiere. "A Functional Partial Semantics for Intensional Logic". In: *Notre Dame Journal of Formal Logic* 33.4 (1992), pp. 517–541. DOI: https://doi.org/10.1305/ndjfl/1093634484.

[228] Shalom Lappin, ed. *The Handbook of Contemporary Semantic Theory*. 2nd. Blackwell Handbooks in Linguistics 3. Blackwell, 1996. ISBN: 0631187529.

[229] Shalom Lappin and Chris Fox. "Type-Theoretic Logic with an Operational Account of Intensionality". In: *Synthese* 192.3 (2015), pp. 563–584. DOI: https://doi.org/10.1007/s11229-013-0390-1.

[230] Richard K. Larson and Peter Ludlow. "Interpreted Logical Forms". In: *Synthese* 95.3 (1993), pp. 305–355. DOI: https://doi.org/10.1007/BF01063877.

[231] Byeong D. Lee. "The Knower Paradox Revisited". In: *Philosophical Studies* 98.2 (2000), pp. 221–232. DOI: https://doi.org/10.1023/A:101832642.

[232] Hannes Leitgeb. "HYPE: A System of Hyperintensional Logic". In: *Journal of Philosophical Logic* 48.2 (2019), pp. 305–405. DOI: https://doi.org/10.1007/s10992-018-9467-0.

[233] François Lepage. "Partial Functions in Type Theory". In: *Notre Dame Journal of Formal Logic* 33.4 (1992), pp. 493–516. DOI: https://doi.org/10.1305/ndjfl/1093634483.

[234] Hector J. Levesque. "A Logic of Explicit and Implicit Belief". In: *Proceedings of the Fourth National Conference on Artificial Intelligence (AAAI-84)*. 1984, pp. 198–202.

[235] David K. Lewis. "General Semantics". In: *Synthese* 22.1–2 (1970), pp. 18–67. DOI: https://doi.org/10.1007/BF00413598.

[236] David K. Lewis. "Languages and Language". In: *Philosophical Papers Volume I*. Oxford University Press, 1983, pp. 163–188. ISBN: 978-0195032048.

[237] Sten Lindström. "Possible Worlds Semantics and the Liar: Reflections on a Problem Posed by Kaplan". In: *The Philosophy of David Kaplan*. Ed. by Joseph Almog and Paolo Leonardi. Oxford University Press, 2009, pp. 93–108. ISBN: 978-0195367881. DOI: https://doi.org/10.1093/acprof:oso/9780195367881.003.0007.

[238] Godehard Link, ed. *One Hundred Years of Russell's Paradox*. de Gruyter Series in Logic and Its Applications, Book 6. De Gruyter, 2004. ISBN: 978-3110174380.

[239] Bernard Linsky. "Logical Types in Some Arguments about Knowability and Belief". In: *New Essays on the Knowability Paradox*. Ed. by Joe Salerno. Oxford University Press, 2009, pp. 163–179. ISBN: 978-0199285495.

[240] Bernard Linsky. *Russell's Metaphysical Logic*. CSLI Publications, 1999. ISBN: 978-1575862095.

[241] Stephen Maitzen. "The Knower Paradox and Epistemic Closure". In: *Synthese* 114.2 (1998), pp. 337–354. DOI: https://doi.org/10.1023/A:1005064624642.

[242] David C. Makinson. "Paradox of the Preface". In: *Analysis* 25.6 (1965), pp. 205–207. DOI: https://doi.org/10.2307/3326519.

[243] Ruth Barcan Marcus. *Modalities: Philosophical Essays*. Oxford University Press, 1993. ISBN: 978-0195096576.

[244] Edwin Mares. "Propositional Functions". In: *The Stanford Encyclopedia of Philosophy*. Ed. by Edward N. Zalta. Fall 2014. 2014. URL: https://plato.stanford.edu/entries/propositional-function.

[245] Per Martin-Löf. *Intuitionistic Type Theory*. Bibliopolis, 1984. ISBN: 978-8870881059.

[246] Per Martin-Löf. "On the Meanings of the Logical Constants and the Justifications of the Logical Laws". In: *Nordic Journal of Philosophical Logic* 1.1 (1996), pp. 11–60.

[247] Per Martin-Löf. "Verificationism Then and Now". In: *Judgement and the Epistemic Foundation of Logic*. Ed. by Maria van der Schaar. Springer, 2012, pp. 3–14. ISBN: 978-94-007-5136-1. DOI: https://doi.org/10.1007/978-94-007-5137-8_1.

[248] Pavel Materna. *Conceptual Systems*. Logos, 2004. ISBN: 3-8325-0636-5.

[249] Thomas McKay and Michael Nelson. "Propositional Attitude Reports". In: *The Stanford Encyclopedia of Philosophy*. Ed. by Edward N. Zalta. Spring 2014. 2014. URL: https://plato.stanford.edu/archives/spr2014/entries/prop-attitude-reports.

[250] Chris Menzel. "Possible Worlds". In: *The Stanford Encyclopedia of Philosophy*. Ed. by Edward N. Zalta. Summer 2014. 2014. URL: http://plato.stanford.edu/archives/sum2014/entries/possible-worlds.

[251] John J.Ch. Meyer. "Epistemic Logic". In: *The Blackwell Guide to Philosophical Logic*. Ed. by Lou Goble. Blackwell, 2001, pp. 183–202. ISBN: 978-0631206934.

[252] John J.Ch. Meyer. "Modal Epistemic and Doxastic Logic". In: *Handbook of Philosophical Logic*. Ed. by Dov M. Gabbay and Franz Guenthner. 2nd. Kluwer Academic Publishers, 2003, pp. 1–38. ISBN: 978-90-481-6431-8. DOI: https://doi.org/10.1007/978-94-017-4524-6_1.

[253] John J.Ch. Meyer and Wiebe van der Hoek. *Epistemic Logic for AI and Computer Science.* Cambridge Tracts in Theoretical Computer Science 41. Cambridge University Press, 1995. ISBN: 0-521-46014-X.

[254] Greg Michaelson. *An Introduction to Functional Programming through Lambda Calculus.* Addison–Wesley, 1989. ISBN: 0201178125.

[255] Eugen Moggi. "The Partial Lambda-Calculus". Ph.D. thesis. University of Edinburgh, 1988.

[256] Friederike Moltmann. "Propositional Attitudes without Propositions". In: *Synthese* 135.1 (2003), pp. 77–118. DOI: https://doi.org/10.1023/A:1022945009188.

[257] Richard Montague. "English as a Formal Language". In: *Linguaggi nella Societa et nella Technica.* Ed. by B. Visentini et al. 1970, pp. 188–221.

[258] Richard Montague. *Formal Philosophy. Selected Papers of Richard Montague edited by R. Thomason.* Yale University Press, 1974. ISBN: 978-0300024128.

[259] Richard Montague. "Logical Necessity, Physical Necessity, Ethics and Quantifiers". In: *Inquiry* 3.1–4 (1960), pp. 259–269. DOI: https://doi.org/10.1080/00201746008601312.

[260] Richard Montague. "On the Nature of Certain Philosophical Entities". In: *The Monist* 53.2 (1969), pp. 159–194. DOI: https://doi.org/10.5840/monist19695327.

[261] Richard Montague. "Syntactical Treatmens of Modality with Corollaries on Reflection Principles and Finite Axiomatizability". In: *Acta Philosophica Phennica* 16 (1963), pp. 153–167.

[262] Richard Montague. "The Proper Treatment of Quantification in Ordinary English". In: *Approaches to Natural Language. Proceedings of the 1970 Stanford Workshop on Grammar and Semantics.* Ed. by K. J. J. Hintikka, J. M. E. Moravcsik, and P. Suppes. D. Reidel, 1973, pp. 221–242. ISBN: 978-90-277-0233-3. DOI: https://doi.org/10.1007/978-94-010-2506-5_10.

[263] Richard Montague. "Universal Grammar". In: *Theoria* 36.3 (1970), pp. 373–398. DOI: https://doi.org/10.1111/j.1755-2567.1970.tb00434.x.

[264] George E. Moore. "Moore's Paradox". In: *Moore: Selected Writings.* Ed. by Thomas Baldwin. Routledge, 1993, pp. 207–212. ISBN: 978-0415098540.

[265] Yiannis N. Moschovakis. "A Logical Calculus of Meaning and Synonymy". In: *Linguistics and Philosophy* 29.1 (2005), pp. 27–89. DOI: https://doi.org/10.1007/s10988-005-6920-7.

[266] Yiannis N. Moschovakis. "Sense and Denotation as Algorithm and Value". In: *Logic Colloquium '90.* Ed. by Jouko Väänanen and Juha Oikkonen. Vol. 2. Lecture Notes in Logic. Association for Symbolic Logic, 1993, pp. 210–249.

[267] Reinhard Muskens. "A Relational Formulation of the Theory of Types". In: *Linguistics and Philosophy* 12.3 (1989), pp. 325–346. DOI: https://doi.org/10.1007/BF00635639.

[268] Reinhard Muskens. "Higher Order Modal Logic". In: *The Handbook of Modal Logic.* Ed. by Patrick Blackburn, Johann F.A.K. van Benthem, and Frank Wolte. Elsevier, 2007, pp. 621–653. ISBN: 978-0-444-51690-9. DOI: https://doi.org/10.1016/S1570-2464(07)80013-9.

[269] Reinhard Muskens. "Intensional Models for the Theory of Types". In: *Journal of Symbolic Logic* 72.1 (2005), pp. 98–118. DOI: https://doi.org/10.2178/jsl/1174668386.

[270] Reinhard Muskens. *Meaning and Partiality.* CSLI, 1995. ISBN: 978-1881526797.

[271] Reinhard Muskens. "Sense and the Computation of Reference". In: *Linguistics and Philosophy* 28.4 (2005), pp. 473–504. DOI: https://doi.org/10.1007/s10988-004-7684-1.

[272] John Myhill. "Problems Arising in the Formalization of Intensional Logic". In: *Logique et Analyse* 1.2 (1958), pp. 74–83.

[273] Rob Nederpelt and Herman Geuvers. *Type Theory and Formal Proof: An Introduction.* Cambridge University Press, 2014. ISBN: 978-1107036505.

[274] Sara Negri, Jan von Plato, and Aarne Ranta. *Structural Proof Theory.* Cambridge University Press, 2001. ISBN: 978-0511527340. DOI: https://doi.org/10.1017/CBO9780511527340.

[275] Bengt Nordström, Kent Petersson, and Jan M. Smith. *Programming in Martin-Löf's Type Theory.* Oxford University Press, 1990. ISBN: 978-0198538141.

[276] Vilém Novák. "Towards Fuzzy Type Theory with Partial Functions". In: *Proceedings of the Conference of the European Society for Fuzzy Logic and Technology.* Ed. by Janusz Kacprzyk et al. Springer, 2017, pp. 25–37. ISBN: 978-3-319-66826-0. DOI: https://doi.org/10.1007/978-3-319-66827-7_3.

[277] Graham Oddie. *Likeness to Truth.* Western Ontario Press, 1986. ISBN: 978-94-009-4658-3. DOI: https://doi.org/10.1007/978-94-009-4658-3.

[278] Graham Oddie and Pavel Tichý. "The Logic of Ability, Freedom and Responsibility". In: *Studia Logica* 41.2–3 (1982), pp. 227–248. DOI: https://doi.org/10.1007/BF00370346.

[279] Mika Oksanen. "The Russell-Kaplan Paradox and other Modal Paradoxes: A New Solution". In: *Nordic Journal of Philosophical Logic* 4.1 (1999), pp. 73–93.

[280] Steven Orey. "Model Theory for the Higher Order Predicate Calculus". In: *Transactions of the American Mathematical Society* 92.1 (1959), pp. 72–84. DOI: https://doi.org/10.2307/1993168.

[281] Francesco Orilia and Gregory Landini. "Truth, Predication and a Family of Contingent Paradoxes". In: *Journal of Philosophical Logic* 48.3 (2019), pp. 113–136. DOI: https://doi.org/10.1007/s10992-018-9480-3.

[282] Naomi Osorio-Kupferblum. "Aboutness". In: *Analysis* 76.4 (2016), pp. 528–546. DOI: https://doi.org/10.1093/analys/anw027.

[283] Gary Ostertag. "Structured Propositions and the Logical Form of Predication". In: *Synthese* 196.4 (2019), pp. 1475–1499. DOI: https://doi.org/10.1007/s11229-017-1420-1.

[284] Eric Pacuit. *Neighborhood Semantics for Modal Logic.* Springer, 2017. ISBN: 978-3-319-67149-9. DOI: https://doi.org/10.1007/978-3-319-67149-9.

[285] Peter Pagin. "A General Argument against Structured Propositions". In: *Synthese* 196.4 (2019), pp. 1501–1528. DOI: https://doi.org/10.1007/s11229-016-1244-4.

[286] Federico Matias Pailos and Lucas Daniel Rosenblatt. "Solving Multimodal Paradoxes". In: *Theoria* 81.3 (2015), pp. 192–210. DOI: https://doi.org/10.1111/theo.12052.

[287] Erik Palmgren. "A Constructive Examination of a Russell-style Ramified Type Theory". In: *Bulletin of Symbolic Logic* 24.1 (2018), pp. 90–106. DOI: https://doi.org/10.1017/bsl.2018.4.

[288] Barbara H. Partee, Alice G.B. ter Meulen, and Robert E. Wall. *Mathematical Methods in Linguistics.* Kluwer Academic Publishers, 1990. ISBN: 90-277-2244-7.

[289] Barbra H. Partee and Herman Hendriks. "Montague Grammar". In: *Handbook of Logic and Language*. Ed. by Johann F.A.K. van Benthem and Alice G.B. ter Meulen. Elsevier,MIT Press, 1997, pp. 5–92. ISBN: 978-0-444-81714-3.

[290] Alexander Paseau. "Fitch's Argument and Typing Knowledge". In: *Notre Dame Journal of Formal Logic* 49.2 (2008), pp. 153–176. DOI: https://doi.org/10.1215/00294527-2008-005.

[291] Alexander Paseau. "How to Type: Reply to Halbach". In: *Analysis* 69.2 (2009), pp. 280–286. DOI: https://doi.org/10.1093/analys/anp016.

[292] Anthony F. Peressini. "Cumulative versus Noncumulative Ramified Types". In: *Notre Dame Journal of Formal Logic* 38.3 (1997), pp. 385–397. DOI: https://doi.org/10.1305/ndjfl/1039700745.

[293] Stanley Peters and Dag Westerståhl. *Quantifiers in Language and Logic*. Clarendon Press, 2006. ISBN: 978-0199291250.

[294] Ivo Pezlar. "Algorithmic Theories of Problems. A Constructive and a Non-Constructive Approach". In: *Logic and Logical Philosophy* 26.4 (2017), pp. 473–508. DOI: https://doi.org/10.12775/LLP.2017.010.

[295] Ivo Pezlar. "Investigations into Transparent Intensional Logic: Rule-based Approach". Ph.D. thesis. Brno: Masarykova univerzita, 2016.

[296] Ivo Pezlar. "On Two Notions of Computation in Transparent Intensional Logic". In: *Axiomathes* 29.62 (2019), pp. 189–205. DOI: https://link.springer.com/article/10.1007/s10516-018-9401-7.

[297] Ivo Pezlar. "Proof-theoretic Semantics and Hyperintensionality". In: *Logique et Analyse* 61.242 (2018), pp. 151–161. DOI: https://doi.org/10.2143/LEA.242.0.3284748.

[298] Ivo Pezlar. "Tichý's Two-Dimensional Conception of Inference". In: *Organon F* 20.Supplementary Issue 2 (2013), pp. 54–65.

[299] Ivo Pezlar and Jiří Raclavský. *Rethinking the Paradox of Logical Omniscience from the Viewpoint of Hyperintensional Logic*. ms. 2016.

[300] Thomas Piecha and Peter Schroeder-Heister, eds. *Advances in Proof-Theoretic Semantics*. Trends in Logic 43. Springer, 2016. ISBN: 978-3-319-22685-9. DOI: https://doi.org/10.1007/978-3-319-22686-6.

[301] Benjamin C. Pierce. *Types and Programming Languages*. MIT Press, 2002. ISBN: 978-0262162098.

[302] Gordon David Plotkin. "LCF Considered as Programming Language". In: *Theoretical Computer Science* 5.3 (1977), pp. 223–255. DOI: https://doi.org/10.1016/0304-3975(77)90044-5.

[303] Carl Pollard. "Agnostic Hyperintensional Semantics". In: *Synthese* 192.3 (2015), pp. 535–562. DOI: https://doi.org/10.1007/s11229-013-0373-2.

[304] Carl Pollard. "Hyperintensions". In: *Journal of Logic and Computation* 18.2 (2008), pp. 257–282. DOI: https://doi.org/10.1093/logcom/exm003.

[305] Dag Prawitz. *Natural Deduction: A Proof-Theoretical Study*. Dover Publications, 2006. ISBN: 978-0486446554.

[306] Graham Priest. *In Contradiction: A Study of the Transconsistent*. 2nd edition. Oxford University Press, 2006. ISBN: 978-0199263301.

[307] Graham Priest. "Intensional Paradoxes". In: *Notre Dame Journal of Formal Logic* 32.2 (1991), pp. 193–211. DOI: https://doi.org/10.1305/ndjfl/1093635745.

[308] Graham Priest. "Paraconsistent Logic". In: *Handbook of Philosophical Logic*. Ed. by Dov Gabbay and Franz Guenthner. 2nd edition. Vol. 6. Kluwer Academic Publishers, 2002, pp. 287–393. ISBN: 978-90-481-6004-4. DOI: https://doi.org/10.1007/978-94-017-0460-1_4.

[309] Graham Priest. "Semantic Closure, Descriptions and Non-Triviality". In: *Journal of Philosophical Logic* 28.6 (1999), pp. 549–558. DOI: https://doi.org/10.1023/A:1004608013532.

[310] Arthur Prior. "On a Family of Paradoxes". In: *Notre Dame Journal of Formal Logic* 2.1 (1961), pp. 16–32. DOI: https://doi.org/10.1305/ndjfl/1093956750.

[311] Riccardo Pucella. "Deductive Algorithmic Knowledge". In: *Journal of Logic Computation* 16.2 (2006), pp. 287–309. DOI: https://doi.org/10.1093/logcom/exi078.

[312] Pavel Pudlák. "The Lengths of Proofs". In: *Handbook of Proof Theory*. Ed. by Samuel R. Buss. Vol. 137. Elsevier, 1998, pp. 547–637. ISBN: 978-0-444-89840-1. DOI: https://doi.org/10.1016/S0049-237X(98)80023-2.

[313] Ruy J G B de Queiroz, Anjolina G de Oliveira, and Dov M Gabbay. *The Functional Interpretation of Logical Deduction*. World Scientific, 2011. ISBN: 978-981-4360-95-1.

[314] Willard V.O. Quine. "On the Theory of Types". In: *Journal of Symbolic Logic* 3.4 (1938), pp. 125–139. DOI: https://doi.org/10.2307/2267776.

[315] Willard V.O. Quine. "Quantifiers and Propositional Attitudes". In: *The Journal of Philosophy* 53.5 (1956), pp. 177–187. DOI: https://doi.org/10.2307/2022451.

[316] Willard V.O. Quine. "Reply to Professor Marcus". In: *Synthese* 13.4 (1962), pp. 323–330. DOI: https://doi.org/10.1007/BF00486630.

[317] Willard V.O. Quine. *Word and Object*. MIT Press, 1960. ISBN: 0-262-67001-1.

[318] Jiří Raclavský. "A Model of Language in a Synchronic and Diachronic Sense". In: *Issues in Philosophy of Language and Linguistic (Lodz Studies in English and General Linguistic 2)*. Ed. by Piotr Stalmaszczyk. Lodz University Press, 2014, pp. 109–123. ISBN: 978-83-7969-402-0.

[319] Jiří Raclavský. "Conceptual Dependence of Verisimilitude Vindicated". In: *Organon F* 15.3 (2008), pp. 369–382.

[320] Jiří Raclavský. "Constructional vs. Denotational Conception of Aboutness". In: *Organon F* 21.2 (2014), pp. 219–236.

[321] Jiří Raclavský. "Executions vs. Constructions". In: *Logica et Methodologica (Anaphora, Logic and Natural Language)*. Univerzita Komenského, 2003, pp. 63–72. ISBN: 978-8022318716.

[322] Jiří Raclavský. "Existential Import and Relations of Categorical and Modal Categorical Statements". In: *Logic and Logical Philosophy* 27.3 (2018), pp. 271–300. DOI: https://doi.org/10.12775/LLP.2017.026.

[323] Jiří Raclavský. "Explicating Truth in Transparent Intensional Logic". In: *Recent Trends in Philosophical Logic*. Ed. by Roberto Ciuni, Heinrich Wansing, and Caroline Willkommen. Springer Verlag, 2014, pp. 167–177. ISBN: 978-3-319-06079-8. DOI: https://doi.org/10.1007/978-3-319-06080-4_12.

[324] Jiří Raclavský. "Explikace a dedukce: od jednoduché k rozvětvené teorii typů [Explication and Deduction: from Simple to Ramified Theory of Types]". In: *Organon F* 20.Supplementary Issue 2 (2013), pp. 37–53.

[325] Jiří Raclavský. "Fitchův paradox poznatelnosti a rozvětvená teorie typů [Fitch's Knowability Paradox and Ramified Theory of Types]". In: *Organon F* 20.Supplementary Issue 1 (2013), pp. 144–165.

[326] Jiří Raclavský. "Is Logical Analysis of Natural Language a Translation?" In: *Philosophy of Language and Linguistics Volume I: The Formal Turn*. Ed. by Piotr Stalmaszczyk. Ontos Verlag, 2010, pp. 229–243. ISBN: 978-3-86838-070-5.

[327] Jiří Raclavský. "Je Tichého logika logikou? (O vztahu logické analýzy a dedukce) [Is Tichý's Logic a Logic? On the Relation of Logical Analysis and Deduction". In: *Filosofický časopis* 60.2 (2012), pp. 245–254.

[328] Jiří Raclavský. *Jména a deskripce: logicko-sémantická zkoumání [Names and Descriptions: Logico-Semantical Considerations]*. Nakladatelství Olomouc, 2009. ISBN: 978-80-7182-277-6.

[329] Jiří Raclavský. "On Interaction of Semantics and Deduction in Transparent Intensional Logic (Is Tichý's Logic a Logic?)" In: *Logic and Logical Philosophy* 23.1 (2014), pp. 57–68. DOI: https://doi.org/10.12775/LLP.2013.035.

[330] Jiří Raclavský. "On Partiality and Tichý's Transparent Intensional Logic". In: *Hungarian Philosophical Review* 54.4 (2010), pp. 120–128.

[331] Jiří Raclavský. "Russellian Typing Knowledge and Fitch's Paradox of Knowability". In: *Aftermath of the Logical Paradise*. Ed. by Jean-Yves Béziau, Alexandre Costa-Leite, and Itala Maria Loffredo D'Ottaviano. 401–423. Colecao CLE, 2017. ISBN: 978-85-86497-36-0.

[332] Jiří Raclavský. "Semantic Concept of Existential Presupposition". In: *Human Affairs* 21.3 (2011), pp. 249–261. DOI: https://doi.org/10.2478/s13374-011-0026-4.

[333] Jiří Raclavský. "Semantic Paradoxes and Transparent Intensional Logic". In: *The Logica Yearbook 2011*. Ed. by Michal Peliš and Vít Punčochář. College Publications, 2012, pp. 239–252. ISBN: 978-1848900714.

[334] Jiří Raclavský. "Tichý's Possible Worlds". In: *Organon F* 21.4 (2014), pp. 471–491.

[335] Jiří Raclavský. "Two Standard and Two Modal Squares of Opposition". In: *The Square of Opposition: A Cornerstone of Thought, Studies in Universal Logic*. Ed. by Jean-Yves Béziau and Gianfranco Basti. Birkhäuser, 2017, pp. 119–142. ISBN: 978-3-319-45061-2. DOI: https://doi.org/10.1007/978-3-319-45062-9_8.

[336] Jiří Raclavský. "Typing Approach to Church-Fitch's Knowability Paradox and its Revenge Form". In: *Prolegomena* 17.1 (2018), pp. 31–49. DOI: https://doi.org/10.26362/20180202.

[337] Jiří Raclavský. "Základy explikace sémantických pojmů [Foundation of Explication of Semantic Concepts]". In: *Organon F* 19.4 (2012), pp. 488–505.

[338] Jiří Raclavský and Petr Kuchyňka. "Conceptual and Derivation Systems". In: *Logic and Logical Philosophy* 20.1–2 (2011), pp. 159–174. DOI: https://doi.org/10.12775/LLP.2011.008.

[339] Jiří Raclavský, Petr Kuchyňka, and Ivo Pezlar. *Transparentní intenzionální logika jako characteristica universalis a calculus ratiocinator. [Transparent Intensional Logic as Characteristica Universalis and Calculus Ratiocinator]*. Masarykova univerzita (Munipress), 2015. ISBN: 978-80-210-7973-1.

[340] Jiří Raclavský and Ivo Pezlar. "Explicitní/implicitní přesvědčení a derivační systémy [Explicit/Implicit Belief and Derivation Systems]". In: *Filosofický časopis* 67.1 (2019), pp. 89–120.

[341] Frank P. Ramsey. "The Foundations of Mathematics". In: *Proceedings of the London Mathematical Society* s2-25.1 (1926), pp. 338–384. DOI: https://doi.org/10.1112/plms/s2-25.1.338.

[342] Aarne Ranta. *Type-Theoretical Grammar*. A Clarendon Press Publication, 1994. ISBN: 978-0198538578.

[343] Veikko Rantala. "Impossible Worlds Semantics and Logical Omniscience". In: *Acta Philosophica Fennica* 35 (1982), pp. 106–115.

[344] Veikko Rantala. "Urn Models: A New Kind of Non-standard Models for First-order Logic". In: *Journal of Philosophical Logic* 4.4 (1975), pp. 455–474. DOI: `https://doi.org/10.1007/BF00558760`.

[345] Mattias Skipper Rasmussen. "Dynamic Epistemic Logic and Logical Omniscience". In: *Logic and Logical Philosophy* 24.3 (2015), pp. 377–379. DOI: `https://doi.org/10.12775/LLP.2015.014`.

[346] William N. Reinhardt. "Necessity Predicates and Operators". In: *Journal of Philosophical Logic* 9.4 (1980), pp. 437–450. DOI: `https://doi.org/10.1007/BF00262865`.

[347] Mark Richard. "Seeking a Centaur, Adoring Adonis: Intensional Transitives and Empty Terms". In: *Midwest Studies In Philosophy* 25.1 (2001), pp. 103–127. DOI: `https://doi.org/10.1111/1475-4975.00041`.

[348] Mark E. Richard. *Propositional Attitudes: An Essay on Thoughts and How We Ascribe Them*. Cambridge University Press, 1990. ISBN: 0-521-38819-8.

[349] Maarten de Rijke, ed. *Advances in Intensional Logic*. Springer, 1997. ISBN: 978-90-481-4897-4. DOI: `https://doi.org/10.1007/978-94-015-8879-9`.

[350] David Ripley. "Structures and Circumstances: Two Ways to Fine-grain Propositions". In: *Synthese* 189.1 (2012), pp. 97–118. DOI: `https://doi.org/10.1007/s11229-012-0100-4`.

[351] Lucas Rosenblatt. "The Knowability Argument and the Syntactic Type-Theoretic Approach". In: *Theoria: An International Journal for Theory, History and Foundations of Science* 29.2 (2014), pp. 201–221. DOI: `https://doi.org/10.1387/theoria.7225`.

[352] Bertrand Russell. *Introduction to Mathematical Philosophy*. George Allen and Unwin, 1919.

[353] Bertrand Russell. "Mathematical Logic as Based on the Theory of Types". In: *American Journal of Mathematics* 30.3 (1908), pp. 222–262. DOI: `https://doi.org/10.2307/2369948`.

[354] Bertrand Russell. "On Denoting". In: *Mind* 14.56 (1905), pp. 479–493.

[355] Bertrand Russell. "The Philosophy of Logical Atomism (Lectures 5-6)". In: *The Monist* 29.2 (1918), pp. 190–222. DOI: `https://doi.org/10.5840/monist19192922`.

[356] Bertrand Russell. *The Principles of Mathematics*. W.W. Norton & Company, 1903.

[357] Bertrand Russell. *The Problems of Philosophy*. Oxford University Press, 1912.

[358] R. Mark Sainsbury. *Paradoxes*. 2nd. Cambridge University Press, 1995. ISBN: 978-1139166775. DOI: `https://doi.org/10.1017/CBO9781139166775`.

[359] Nathan Salmon. "Reference and Information Content: Names and Description". In: *Handbook of Philosophical Logic, vol. 10*. Ed. by Dov M. Gabbay and Franz Guenthner. 2nd. Kluwer Academic Publishers, 2003, pp. 39–85. ISBN: 978-90-481-6431-8. DOI: `https://doi.org/10.1007/978-94-017-4524-6_2`.

[360] Nathan U. Salmon. *Frege's Puzzle*. Bradford Books, 1986. ISBN: 0262192462.

[361] Giorgio Sbardolini. "On Hierarchical Propositions". In: *Journal of Philosophical Logic* unassigned (2019). DOI: `https://doi.org/10.1007/s10992-019-09509-9`.

[362] Burkhard C. Schipper. "Awareness". In: *Handbook of Epistemic Logic*. Ed. by Hans van Ditmarsch et al. College Publications, 2015, pp. 77–146. ISBN: 978-1848901582.

[363] Moses Schönfinkel. "On the Building Blocks of Mathematical Logic". In: *From Frege To Gödel: A Source Book in Mathematical Logic, 1879–1931*. Ed. by Jean van Heijenoort. Harvard University Press, 1967, pp. 355–366. ISBN: 0-674-32449-8.

[364] Peter Schroeder-Heister. "Proof-Theoretic Semantics". In: *The Stanford Encyclopedia of Philosophy*. Ed. by Edward N. Zalta. Spring 2018. 2018. URL: https://plato.stanford.edu/archives/spr2018/entries/proof-theoretic-semantics.

[365] Peter Schroeder-Heister. "Rules of Definitional Reflection". In: *Proceedings of the 8th Annual IEEE Symposium on Logic in Computer Science*. IEEE Press, 1993, pp. 222–232. ISBN: 0-8186-3140-6. DOI: 10.1109/LICS.1993.287585.

[366] Eric Schwitzgebel. "Belief". In: *The Stanford Encyclopedia of Philosophy*. Ed. by Edward N. Zalta. Summer 2015. Metaphysics Research Lab, Stanford University, 2015. URL: https://plato.stanford.edu/archives/sum2015/entries/belief.

[367] Dana Scott. "Advice on Modal Logic". In: *Philosophical Problems in Logic*. Ed. by Karel Lambert. Springer, 1970, pp. 143–173. DOI: https://doi.org/10.1007/978-94-010-3272-8_7.

[368] Dana Scott and Christopher Strachey. *Toward a Mathematical Semantics for Computer Languages*. Oxford Programming Research Group Technical Monograph. PRG-6, 1971.

[369] Igor Sedlár. "An Outline of a Substructural Model of BTA Belief". In: *Organon F* 20.Supplementary Issue 2 (2013), pp. 160–170.

[370] Igor Sedlár. "Hyperintensional Logics for Everyone". In: *Synthese* unassigned (2019). DOI: https://doi.org/10.1007/s11229-018-02076-7.

[371] Gila Sher. *The Bounds of Logic: A Generalized Viewpoint*. The MIT Press, Bradford Books, 1991. ISBN: 978-0262193115.

[372] Yaroslav Shramko and Heinrich Wansing. *Truth and Falsehood: An Inquiry into Generalized Logical Values*. Springer, 2012. ISBN: 978-94-007-0906-5. DOI: https://doi.org/10.1007/978-94-007-0907-2.

[373] Yaroslav Shramko and Heinrich Wansing. "Truth Values". In: *The Stanford Encyclopedia of Philosophy*. Ed. by Edward N. Zalta. Spring 2018. 2018. URL: https://plato.stanford.edu/archives/spr2018/entries/truth-values.

[374] Mattias Skipper and Jens Christian Bjerring. "Hyperintensional Semantics: a Fregean Approach". In: *Synthese* unassigned (2018). DOI: https://doi.org/10.1007/s11229-018-01900-4.

[375] Sonja Smets and Anthia Solaki. "The Effort of Reasoning: Modelling the Inference Steps of Boundedly Rational Agents". In: *Logic, Language, Information, and Computation: Proceedings of 25th International Workshop*. Ed. by Lawrence S. Moss, Ruy de Queiroz, and Maricarmen Martinez. Springer, 2018, pp. 307–324. ISBN: 978-3-662-57668-7. DOI: https://doi.org/10.1007/978-3-662-57669-4_18.

[376] Sonja Smets and Fernando Velázquez-Quesada. "Philosophical Aspects of Multi-Modal Logic". In: *The Stanford Encyclopedia of Philosophy*. Ed. by Edward N. Zalta. Summer 2019 Edition. URL: https://plato.stanford.edu/archives/sum2019/entries/phil-multimodallogic/.

[377] Craig Smoryński. *Self-Reference and Modal Logic*. Springer, 1985. ISBN: 978-0-387-96209-2. DOI: https://doi.org/10.1007/978-1-4613-8601-8.

[378] Scott Soames. *Beyond Rigidity: The Unfinished Agenda of Naming and Necessity*. Oxford University Press, 2002. ISBN: 978-0195145298.

[379] Scott Soames. "Direct Reference and Propositional Attitudes". In: *Propositional Attitudes: The Role of Content in Logic, Language, and Mind.* Ed. by C. Anthony Anderson and Joseph Owens. Center for the Study of Language and Information, 1990, pp. 393–419. ISBN: 978-0937073506.

[380] Scott Soames. "Direct Reference, Propositional Attitudes, and Semantic Content". In: *Philosophical Topics* 15.1 (1987), pp. 47–87. DOI: https://doi.org/10.5840/philtopics198715112.

[381] Scott Soames. "Presupposition". In: *Handbook of Philosophical Logic, vol. 4.* Ed. by Dov Gabbay and Franz Guenthner. Kluwer Academic Publishers, 1989, pp. 553–616. ISBN: 978-94-010-7021-8. DOI: https://doi.org/10.1007/978-94-009-1171-0_9.

[382] Scott Soames. *What is Meaning?* Princeton University Press, 2010. ISBN: 978-0-691-14640-9.

[383] Scott Soames. "Why Propositions Cannot be Sets of Truth-supporting Circumstances". In: *Journal of Philosophical Logic* 37.3 (2008), pp. 267–276. DOI: https://doi.org/10.1007/s10992-007-9069-8.

[384] Morten Heine Sørensen and Paweł Urzyczyn. *Lectures on the Curry-Howard Isomorphism.* Elsevier, 1998. ISBN: 978-0-444-52077-7.

[385] Roy Sorensen. "Epistemic Paradoxes". In: *The Stanford Encyclopedia of Philosophy.* Ed. by Edward N. Zalta. Summer 2018 Edition. 2018. URL: https://plato.stanford.edu/archives/sum2018/entries/epistemic-paradoxes.

[386] Ernst Sosa. "Propositional Attitudes *de Dicto* and *de Re*: Rejoinder to Hintikka". In: *Journal of Philosophy* 68.16 (1971), pp. 498–501. DOI: https://doi.org/10.2307/2024849.

[387] Robert Stalnaker. "The Problem of Logical Omniscience I." In: *Synthese* 89.3 (1991), pp. 425–440. DOI: https://doi.org/10.1007/BF00413506.

[388] Robert Stalnaker. *Ways a World Might Be: Metaphysical and Anti-metaphysical Essays.* Oxford University Press, 2003. ISBN: 978-0199251483. DOI: https://doi.org/10.1093/0199251487.001.0001.

[389] Robert C. Stalnaker. "Propositions". In: *Issues in the the Philosophy of Language: Colloquium Proceedings.* Ed. by Alfred F. MacKay nad Daniel D. Merill. Yale University Press, 1976, pp. 79–91. ISBN: 978-0300018288.

[390] Robert C. Stalnaker. "Propositions". In: *The Philosophy of Language.* Ed. by A. P. Martinich. Oxford University Press, 1985, pp. 373–380. ISBN: 978-0195035537.

[391] Matthias Steup. "Epistemology". In: *The Stanford Encyclopedia of Philosophy.* Ed. by Edward N. Zalta. Summer 2018. 2018. URL: https://plato.stanford.edu/archives/sum2018/entries/epistemology.

[392] Peter Frederick Strawson. "Identifying Reference and Truth-values". In: *Theoria* 30.2 (1964), pp. 96–118. DOI: https://doi.org/10.1111/j.1755-2567.1964.tb00404.x.

[393] Peter Frederick Strawson. "On Referring". In: *Mind* 59.235 (1950), pp. 320–344.

[394] Göran Sundholm. "Constructions, Proofs and the Meaning of Logical Constants". In: *Journal of Philosophical Logic* 12.2 (1983), pp. 151–172. DOI: https://doi.org/10.1007/BF00247187.

[395] Göran Sundholm. "Proof Theory and Meaning". In: *Handbook of Philosophical Logic.* Ed. by Dov Gabbay and Franz Guenthner. Vol. 9. Springer, 2002, pp. 165–198. ISBN: 978-90-481-6055-6. DOI: https://doi.org/10.1007/978-94-017-0464-9_3.

[396] Göran Sundholm. "Systems of Deduction". In: *Handbook of Philosophical Logic.* Ed. by Dov M. Gabbay and Franz Guenther. Kluwer, 1983, pp. 133–188. ISBN: 978-94-009-7068-7. DOI: https://doi.org/10.1007/978-94-009-7066-3_2.

[397] Koji Tanaka et al., eds. *Paraconsistency: Logic and Applications*. Springer, 2014. ISBN: 978-94-007-4437-0. DOI: https://doi.org/10.1007/978-94-007-4438-7.

[398] Alfred Tarski. "The Concept of Truth in Formalized Languages". In: *Logic, Semantics and Metamathematics*. Ed. by Joseph Henry Woodger. Clarendon Press, 1956, pp. 152–278.

[399] Coq Development Team. *Coq Manuals*. 2018. URL: https://coq.inria.fr/distrib/current/refman.

[400] HOL Development Team. *HOL manuals*. 2018. URL: https://hol-theorem-prover.org.

[401] Isabelle Development Team. *Isabelle Documentation*. 2018. URL: https://www.cl.cam.ac.uk/research/hvg/Isabelle/documentation.html.

[402] Nuprl Development Team. *Nuprl Project Website*. 2018. URL: http://www.nuprl.org.

[403] Neil Tennant. *Natural Logic*. Edinburgh University Press, 1978. ISBN: 978-0852243473.

[404] Elias Thijsse. "Combining Partial and Classical Semantics". In: *Partiality, Modality, and Nonmonotonicity*. Ed. by Patrick Doherty. CSLI, 1996, pp. 223–249. ISBN: 978-1-57586-030-9.

[405] Richmond H. Thomason. "A Model Theory for Propositional Attitudes". In: *Linguistics and Philosophy* 4.1 (1980), pp. 47–70. DOI: https://doi.org/10.1007/BF00351813.

[406] Richmond H. Thomason. "A Note on Syntactical Treatments of Modality". In: *Synthese* 44.3 (1980), pp. 391–395. DOI: https://doi.org/10.1007/BF00413468.

[407] Richmond H. Thomason. "Indirect Discourse is not Quotational". In: *The Monist* 60.3 (1977), pp. 340–354. DOI: https://doi.org/10.5840/monist19776039.

[408] Richmond H. Thomason. "Motivating Ramified Type Theory". In: *Properties, Types and Meaning Vol. I*. Ed. by Gennaro Chierchia, Barbara H. Partee, and Raymond Turner. Kluwer, 1989, pp. 47–62. ISBN: 1-55608-067-0.

[409] Richmond H. Thomason. "Paradoxes and Semantic Representation". In: *Theoretical Aspects of Reasoning About Knowledge, Proceedings of the 1986 Conference*. Ed. by Joseph Y. Halpern. Morgan Kaufman, 1986, pp. 225–239. ISBN: 978-0934613040. DOI: https://doi.org/10.1016/B978-0-934613-04-0.50018-1.

[410] Pavel Tichý. "An Approach to Intensional Analysis". In: *Noûs* 5.3 (1971), pp. 273–297. DOI: https://doi.org/10.2307/2214668.

[411] Pavel Tichý. "Constructions". In: *Philosophy of Science* 53.4 (1986), pp. 514–534. DOI: https://doi.org/10.1086/289338.

[412] Pavel Tichý. "Cracking the Natural Language Code". In: *From the Logical Point of View* 3 (1994), pp. 6–19.

[413] Pavel Tichý. "Einzeldinge als Amtsinhaber". In: *Zeitschrift für Semiotik* 9.1–2 (1987), pp. 13–50.

[414] Pavel Tichý. "Existence and God". In: *The Journal of Philosophy* 76.8 (1979), pp. 403–420. DOI: https://doi.org/10.2307/2025409.

[415] Pavel Tichý. "Foundations of Partial Type Theory". In: *Reports on Mathematical Logic* 14 (1982), pp. 57–72.

[416] Pavel Tichý. "Indiscernibility of Identicals". In: *Studia Logica* 45.3 (1986), pp. 251–273. DOI: https://doi.org/10.1007/BF00375897.

[417] Pavel Tichý. "Intensions in Terms of Turing Machines". In: *Studia Logica* 24.1 (1969), pp. 7–21. DOI: `https://doi.org/10.1007/BF02134290`.

[418] Pavel Tichý. "Introduction to Intensional Logic". unpublished ms. 1976.

[419] Pavel Tichý. *Pavel Tichý's Collected Papers in Logic and Philosophy.* Ed. by Vladimír Svoboda, Bjørn Jespersen, and Colin Cheyne. The University of Otago Press and Filosofia, 2004. ISBN: 978-80-7007-189-2.

[420] Pavel Tichý. "Smysl a procedura [Sense and Procedure]". In: *Filosofický časopis* 16 (1968), pp. 222–232.

[421] Pavel Tichý. "The Analysis of Natural Language". In: *From the Logical Point of View* 3.2 (1994), pp. 42–80.

[422] Pavel Tichý. *The Foundations of Frege's Logic.* Walter de Gruyter, 1988. ISBN: 978-3110116687.

[423] Pavel Tichý. "The Logic of Temporal Discourse". In: *Linguistics and Philosophy* 3.3 (1980), pp. 343–369. DOI: `https://doi.org/10.1007/BF00401690`.

[424] Pavel Tichý. "Two Kinds of Intensional Logic". In: *Epistemologia* 1.1 (1978), pp. 143–164.

[425] Pavel Tichý and Graham Oddie. "Ability and Freedom". In: *American Philosophical Quarterly* 20.2 (1983), pp. 135–147.

[426] Pavel Tichý and Jindra Tichý. "On Inference". In: *The Logica Yearbook 1998.* Ed. by Timothy Childers. Filosofia, 1999, pp. 73–85. ISBN: 978-8070071236.

[427] Luca Tranchini. "Proof-Theoretic Semantics, Paradoxes, and the Distinction between Sense and Denotation". In: *Journal of Logic and Computation* 26.2 (2016), pp. 495–512. DOI: `https://doi.org/10.1093/logcom/exu028`.

[428] Anne Sjerp Troelstra and Helmut Schwichtenberg. *Basic Proof Theory.* Cambridge University Press, 1996. ISBN: 0-521-57223-1.

[429] Adam Trybus. "Leon Chwistek, The Principles of the Pure Type Theory (1922), translated by Adam Trybus with an Introductory Note by Bernard Linsky". In: *Philosophia Naturalis* 33.4 (2012), pp. 329–352. DOI: `https://doi.org/10.1080/01445340.2012.695104`.

[430] Dustin Tucker. "Montagovian Paradoxes and Hyperintensional Content". In: *Studia Logica* 105.1 (2017), pp. 153–171. DOI: `https://doi.org/10.1007/s11225-016-9685-9`.

[431] Dustin Tucker. "Paradoxes and the Limits of Theorizing about Propositional Attitudes". In: *Synthese* (2018), pp. 1–20. DOI: `https://doi.org/10.1007/s11229-018-01902-2`.

[432] Dustin Tucker and Richmond H. Thomason. "Paradoxes of Intensionality". In: *Review of Symbolic Logic* 4.3 (2011), pp. 394–411. DOI: `https://doi.org/10.1017/S1755020311000128`.

[433] Raymond Turner. "Properties, Propositions and Semantic Theory". In: *Computational Linguistic and Formal Semantics.* Ed. by Michael Rosner and Roderick Johnson. Cambridge University Press, 1992, pp. 159–180. ISBN: 0-521-42988-9.

[434] Raymond Turner. *Truth and Modality for Knowledge Representation.* Cambridge University Press, 1990. ISBN: 978-0273031864.

[435] Raymond Turner. "Types". In: *Handbook of Logic and Language.* Ed. by Johann F.A.K van Benthem nad Alice G.B. ter Meulen. Elsevier, 1997, pp. 535–586. ISBN: 0-262-22053-9.

[436] Thomas Tymoczko. "An Unsolved Puzzle about Knowledge". In: *The Philosophical Quaterly* 34.137 (1984), pp. 437–458. DOI: https://doi.org/10.2307/2219063.

[437] Gabriel Uzquiano. "The Paradox of the Knower without Epistemic Closure?" In: *Mind* 113.449 (2004), pp. 95–107.

[438] Jouko Väänänen. "Second-order and Higher-order Logic". In: *The Stanford Encyclopedia of Philosophy*. Ed. by Edward N. Zalta. Fall 2019. 2019. URL: https://plato.stanford.edu/archives/fall2019/entries/logic-higher-order/.

[439] Fernando R. Velázquez-Quesada. "Explicit and Implicit Knowledge in Neighbourhood Models". In: *Logic, Rationality, and Interaction - 4th International Workshop, LORI 2013, Hangzhou, China, October 9-12, 2013, Proceedings*. Ed. by Davide Grossi, Olivier Roy, and Huaxin Huang. Springer, 2013, pp. 239–252. ISBN: 978-3-642-40947-9. DOI: https://doi.org/10.1007/978-3-642-40948-6_19.

[440] Albert Visser. "Semantics and the Liar Paradox". In: *Handbook of Philosophical Logic, vol. 4*. Ed. by Dov Gabbay and Franz Guenthner. Kluwer Academic Publishers, 1989, pp. 617–706. ISBN: 978-94-010-7021-8. DOI: https://doi.org/10.1007/978-94-009-1171-0_10.

[441] The Univalent Foundations Program [Vladimir Voevodsky]. *Homotopy Type Theory: Univalent Foundations of Mathematics*. Princeton: Institute for Advanced Study, 2013. URL: https://homotopytypetheory.org/book.

[442] Heinrich Wansing. "A General Possible Worlds Framework for Reasoning about Knowledge and Belief". In: *Studia Logica* 49.4 (1990), pp. 523–539. DOI: https://doi.org/10.1007/BF00370163.

[443] Heinrich Wansing. "The Idea of a Proof-Theoretic Semantics and the Meaning of the Logical Operations". In: *Studia Logica* 64.1 (2000), pp. 3–20. DOI: https://doi.org/10.1023/A:1005217827758.

[444] Dag Westerståhl. "Quantifiers in Formal and Natural Languages". In: *Handbook of Philosophical Logic, vol. 4*. Ed. by Dov Gabbay and Franz Guenthner. Kluwer Academic Publishers, 1989, pp. 1–131. ISBN: 978-94-010-7021-8. DOI: https://doi.org/10.1007/978-94-009-1171-0_1.

[445] Alfred N. Whitehead and Bertrand Russell. *Principia Mathematica*. Cambridge University Press, 1910–1913.

[446] Alfred N. Whitehead and Bertrand Russell. *Principia Mathematica*. 2nd. Cambridge University Press, 1927. ISBN: 978-0521067911.

[447] Bruno Whittle. "Hierarchical Propositions". In: *Journal of Philosophical Logic* 46.2 (2017), pp. 215–231. DOI: https://doi.org/10.1007/s10992-016-9399-5.

[448] Bartosz Wieckowski. "Constructive Belief Reports". In: *Synthese* 192.3 (2015), pp. 603–633. DOI: https://doi.org/10.1007/s11229-014-0540-0.

[449] Timothy Williamson. *Knowledge and its Limits*. Oxford University Press, 2000. ISBN: 978-0198250432.

[450] Timothy Williamson. *Modal Logic as Metaphysics*. Oxford University Press, 2013. ISBN: 978-0199552078.

[451] Andrzej Wiśniewski. "Propositions, Possible Worlds, and Recursion". In: *Logic and Logical Philosophy* 20.1-2 (2011), pp. 73–79. DOI: https://doi.org/10.12775/LLP.2011.004.

[452] Edward N. Zalta. "A Comparison of Two Intensional Logics". In: *Linguistics and Philosophy* 11.1 (1988), pp. 59–89. DOI: https://doi.org/10.1007/BF00635757.

[453] Edward N. Zalta. *Intensional Logic and the Metaphysics of Intentionality*. The MIT Press, 1988. ISBN: 978-0262519526.

[454] Thomas E. Zimmerman. "Meaning Postulates and the Model-Theoretic Approach to Natural Language Semantics". In: *Linguistics and Philosophy* 22.5 (2006), pp. 529–561. DOI: https://doi.org/10.1023/A:1005409607329.

[455] Marián Zouhar. "Linguistic Knowledge, Semantics, and Ideal Speakers (Some Remarks on the Methodology of Natural Language Semantics)". In: *Epistemologia* 33.1 (2009), pp. 41–64.

Index of Names

Index of Names

Index of Subject

Index of Subject

Index of Main Symbols and Abbreviations

www.ingramcontent.com/pod-product-compliance
Lightning Source LLC
Chambersburg PA
CBHW062158080426
42734CB00010B/1742
* 9 7 8 1 8 4 8 9 0 3 3 4 0 *